ALLE ZEIT WACH
1842

Ulrich Eisele

Introduction to Polymer Physics

With 148 Figures

Springer-Verlag Berlin Heidelberg New York
London Paris Tokyo Hong Kong

Ulrich Eisele
Alfred-Kubin-Straße 13
D-5090 Leverkusen 1
FRG

Translated from the German by Stephen D. Pask
Goethestraße 57D, D-4047 Dormagen 1, FRG

ISBN 3-540-50777-9 Springer-Verlag Berlin Heidelberg New York
ISBN 0-387-50777-9 Springer-Verlag New York Berlin Heidelberg

Library of Congress Cataloging-in-Publication Data
Eisele, Ulrich, 1938– Introduction to polymer physics / Ulrich Eisele. p. cm.
"Translated from the German by Stephen D. Pask" – – T.p. verso.
ISBN 0-387-50777-9 (U.S.) 1. Polymers. I. Title.
QD381.8.E35 1990 547.7 – – dc20 90–9490 CIP

Printed in the United States of America

Typesetting: Elsner & Behrens GmbH, Oftersheim
2152/3140-543210 – Printed on acid-free paper

Foreword

This book is an expanded version of a course of lectures on polymer physics which I have been delivering for a number of years to advanced students at the German Institute of Rubber Technology in Hannover. A large portion of the book is therefore devoted to the physics of elastomers, whose position in polymer physics, notably in the development of theoretical concepts for describing structure-property correlations, is paramount.

The development of high-performance polymeric materials, which is continuing at breathtaking speed, would be inconceivable without simultaneous research by polymer physicists, whose activity can be seen in the formulation of new fundamental model concepts and use of modern methods of physical measurement. As a purposely concise introduction to polymer physics, the book should help the newcomer to get his bearings in this field without finding himself encumbered by excessive theoretical ballast. It is addressed also to physicists, chemists and engineers engaged throughout the chemical industry in research or product development, many of whom would like to be better acquainted with the overriding aims of polymer physics: the unravelling of fundamental relationships between chemical structure, physical supramolecular structure and the technological properties of modern polymeric materials and the utilization of the resulting knowledge in the common quest by members of all the relevant desciplines for new and still more "capable" polymeric materials.

As a researcher in the chemical industry I have the good fortune to be closely concerned with basic practical problems and am privileged to work with colleagues from other disciplines in attempting to solve them. This, I think, has broadened my field of vision and turned my thinking in new directions. In this connection I would like to thank especially my physicist colleagues at Bayer AG, Dr. H. Hespe and Dr. L. Morbitzer, who for many years have helped me very much in critical discussions and by placing results of their own scientific work at my disposal.

I would also like to thank my former teacher, Prof. Dr. W. Pechhold, of Ulm University, who enabled me to present his pioneering theoretical concept of the meander model of polymers – at least in its fundamentals – and has helped me with this task in every possible way.

My thanks are also due to Dr. H. K. Müller for his critical examination of the typescript and many useful suggestions.

I would like to thank, finally, my wife and my assistant Mrs. Franz, both of whom helped in many ways with the preparation of the typescript.

Leverkusen, January 1990 Ulrich Eisele

Table of Contents

Part I The Mechanics of Linear Deformation of Polymers

1	Object and Aims of Polymer Physics	3
2	Mechanical Relaxation in Polymers	5
2.1	Basic Continuum Mechanics	5
2.1.1	Stress and Strain Tensors	5
2.1.2	Basic Laws of Continuum Mechanics	5
2.2	Relaxation and Creep Experiments on Polymers	8
2.2.1	Creep Experiment	8
2.2.2	Relaxation Experiment	9
2.2.3	Basic Law for Relaxation and Creep	10
2.3	Dynamic Relaxation Experiments	11
2.4	Technical Measures for Damping	12
2.4.1	Energy Dissipation Under Defined Load Conditions	12
2.4.2	Rebound Elasticity	15
3	Simple Phenomenological Models	17
3.1	Maxwell's Model	17
3.2	Kelvin-Voigt Model	18
3.3	Relaxation and Retardation Spectra	20
3.4	Approximate Determination of Relaxation Spectra	22
3.4.1	Method of Schwarzl and Stavermann	22
3.4.2	Method According to Ferry and Williams	24
4	Molecular Models of Relaxation Behavior	26
4.1	Simple Jump Model	26
4.2	Change of Position in Terms of a Potential Model	27
4.3	Viscosity in Terms of the Simple Jump Model	29
4.4	Determining the Energy of Activation by Experiment	30
4.5	Kink Model	32
5	Glass Transition	35
5.1	Thermodynamic Description	35
5.2	Free Volume Theory	36
5.3	Williams, Landel and Ferry Relationship	38
5.4	Time-Temperature Superposition Principle	40
5.5	Increment Method for the Determination of the Glass Transition Temperature	43

5.6 Glass Transitions of Copolymers 45
5.7 Dependence of T_g on Molar Mass 47
5.8 Empirical Correlations Between Molecular Parameters and Glass Transition Temperatures 48
5.9 Plasticizer 52
5.10 Crosslinking 53
5.11 Fillers 55

6 Flow and Rubber Elasticity in Polymer Melts 56
6.1 Flow as a Relaxation Process 56
6.2 Structural Models for Polymer Melts 57
6.3 Bueche-Rouse Model 59
6.4 Rouse Theory of Flow in Low Molar Mass Polymer Melts 60
6.5 Extension of the Rouse Theory to Large Molar Mass and Crosslinked Polymer Melts 65
6.6 Relaxation Processes According to the Meander Model 66
6.6.1 Rubber Elasticity 66
6.6.2 Flow 69
6.6.3 Glass Transition Process 69
6.7 Non-Newtonian Viscosity and the Behavior of Polymers During Processing 72

Part II Crystallization and Melting of Polymers

7 Crystallization Behavior 79
7.1 Polymer Crystals and Growth Forms 79
7.2 Crystalline Structures in Stretched Polymers 84
7.3 Nucleation 86
7.4 Crystal Growth 87

8 Melting Behavior 91
8.1 Equilibrium Thermodynamics 91
8.2 Influence of Crystallite Size 92
8.3 Entropy Effects 93
8.3.1 Stress Crystallization 93
8.3.2 Entropy of Mixing 95

Part III Non-linear Deformation Behavior of Polymers

9 Mechanism of Deformation of Thermoplastics and Multi-component Systems 99
9.1 Terminology 99
9.2 Crazing 100
9.3 Shear Deformation 104
9.4 Deformation Mechanisms in Partially Crystalline Thermoplastics 105

10 Rubber Elasticity of Covalently Crosslinked Elastomers 107
10.1 Thermodynamics of Rubber Elasticity 107

10.2 Statistics of the Segment Model 111
10.3 Statistics of Chains with Free Rotation Around Their Bond Angles 113
10.4 Statistics of a Covalent Chain with Hindered Rotation Around the Bonds 115
10.5 Statistical Theory of Rubber Elasticity 117
10.6 Stress-Strain Relationships for Different Types of Applied Stress. 122
10.6.1 Uniaxial Tension or Compression 122
10.6.2 Biaxial Elongation 122
10.6.3 Simple Shear 123
10.7 Phantom Networks 124
10.8 Mooney-Rivlin Theory 125
10.9 Non-Gaussian Chain Statistics and Network Theory 130
10.10 Van der Waals Theory of Networks 133
10.11 Photoelastic Properties of Elastomers 135

11 Tear Formation and Propagation in Elastomers 138
11.1 Concept of Tearing Energy According to Rivlin 138
11.1.1 Trousers Test Piece 139
11.1.2 Tensile Strip with a Small Cut 140
11.1.3 "Pure shear" Test Piece 141
11.2 Elastic Energy Density in an Elastomer 142
11.3 Fatigue Crack Propagation Under Dynamic Load 143

12 Deformation Behavior of Thermoplastic Elastomers 147
12.1 Structural Principles 147
12.2 Polyurethane Elastomers 148
12.3 Block Copolymers 149
12.4 Thermoplastic Elastomers Based on Polymer Mixtures 152
12.5 Tension Set 154

Part IV Mixing and Swelling of Polymers

13 Compatibility of Polymers 159
13.1 Basic Theoretical Considerations 159
13.2 Flory-Huggins Theory 161
13.3 Development of the Flory-Huggins Theory to a Description of Polymer Mixtures 164
13.4 Solubility Parameter 164
13.5 Experimental Methods for Determining Miscibility 167

14 Network Swelling 173

15 Environmental Stress Cracking of Polymeric Materials 177

Part I
The Mechanics of Linear Deformation of Polymers

1 Object and Aims of Polymer Physics

The substitution not only of metals but also of glass, wood, paper and leather by high-quality, high-performance synthetic materials is gathering momentum. The continuation of this substitution process has been further assured by the introduction of blend technology, in particular the blending of thermoplastics, which has led to the development of "High-Tech" polymeric materials. It is doubtful whether these material developments could have been made without the fundamental information derived by a consideration of polymer physics. One of the most important tasks of polymer physics involves, from investigation of molecular mobility and deformation, phase transitions, molecular interactions and the resulting supra-molecular structures, the development of an understanding, from a molecular standpoint, of the physical and technological properties of polymeric materials. Such an understanding is a prerequisite for any systematic modification or optimization, in terms of the continually increasing industrial requirements of existing materials and, indeed, for a programmed development of new materials. Such developments require a continuous feed-back between the polymer physicist and, not only, the synthetic and analytical chemists and the development technologists, but also between the physicist and the manufacturing and process technologists and the constructors and designers of plastic articles.

Until now, no unified, all-embracing theoretical basis has been developed which can account for all the physical properties of polymeric materials in terms of the chemical structure of the individual molecules. Nevertheless, there are a number of theories which can be applied quantitatively to approximate some polymer specific phenomena: Entropy-elasticity, viscosity, glassy solidification, particular deformation and failure mechanisms, mixing behavior, crystallization and melting. A standardization of all phenomena in one, consistent theory requires a knowledge of the form, the force-field and the rotational-potential (i.e. the inter- and intramolecular potentials) of the molecules which constitute the polymeric material. Although such data is becoming increasingly available due to more sophisticated measuring techniques and optimized, computer-aided calculation procedures, the fundamental difficulties associated with the development of an all-encompassing theory concerning the properties of polymeric solids and melts from a molecular standpoint should not be underestimated. The sum of the macromolecules making up a polymeric material represent a multi-particle system in which the individual molecules interact with one another. Since, for most macromolecules, a strong covalent bond between the monomer units and only limited rotation about these bonds are characteristic, the undoubtedly viable model, developed for dilute solutions, of statistically coiled chains has also been applied to the amorphous solid state. It is not difficult to appreciate the immense problems which arise when one attempts to quantitatively allow for the interaction of each and every molecule in an amorphous solid composed of statistically disordered macromolecules: Theoretical polymer physics has, however, not capitulated. A number of promising approaches have been developed which, because of the approximations and particular computer-aided simulation-techniques involved, or because of the unconventional nature of the models proposed for the condensed polymer phase (bundle model), can only approximate, rather than quantitatively describe, the properties of polymeric materials in terms of their molecular structures. In this book such theories will only be summarily treated. Naturally, such attempts are the subject of ongoing, long-term research. For the polymer physicist working with the present state of the art, on industrially related problems concerning technologically optimised materials, which are often highly complex multi-phase or reinforced systems, the existing theories are not suitable as a basis for systematic developments. At the present time the well established semi-empirical approaches still provide the basis for the polymer physicist working in industry. Thus, by systematic variation of the relevant polymer specific parameters the property profile of a material is measured and trends ascertained. From a knowledge of the particular effects of long-chain molecules in such cases it is often possible, with the aid of suitable molecular models, to develop a rational basis for limited structure-property relationships. The polymer physics associated with this approach is the subject of this treatise.

2 Mechanical Relaxation in Polymers

2.1 Basic Continuum Mechanics [1–3]

2.1.1 Stress and Strain Tensors

The state of an ideal elastic body under stress and strain due to the influence of a load can be described by corresponding tensors. The components of the strain tensor ε_{ii} and γ_{ik} determine the relative change in the dimensions and angles of a small, cubic volume element. Such an element is imagined to be extracted from within the body under load. In a similar manner, the components of the stress tensor σ_{ii} and τ_{ik} can be used to determine the forces operating on the surfaces of the imaginary cube. The usual matrix formulation for tensors leads to the following expressions:

$$\varepsilon = \begin{pmatrix} \varepsilon_{11} & \gamma_{12} & \gamma_{13} \\ \gamma_{21} & \varepsilon_{22} & \gamma_{23} \\ \gamma_{31} & \gamma_{32} & \varepsilon_{33} \end{pmatrix} \qquad \sigma = \begin{pmatrix} \sigma_{11} & \tau_{12} & \tau_{13} \\ \tau_{21} & \sigma_{22} & \tau_{23} \\ \tau_{31} & \tau_{32} & \sigma_{33} \end{pmatrix}$$

The stress component τ_{ik}, which is a shear stress, corresponds to a force operating in the k-direction on the surface of the cube with a normal in the i-direction. The normal stresses σ_{ii} are considered as positive for tension and negative for compression. Stress and strain tensors are symmetrical tensors having six independent coefficients and can be represented by a special second order surface – an ellipsoid. If the system of coordinates (i, k, l) is centered on the main axis of the ellipsoid, the mixed tensor components cancel. Using this system of coordinates the stress/strain state of the material can be described using three coefficients corresponding to the diagonals of the coefficient matrix. These three coefficients are accordingly called the main stresses or strains.

2.1.2 Basic Laws of Continuum Mechanics

Hooke's law describes the relationship between the stress and strain tensors. Each of these quantities has six components so that, in the most general case, each component of the stress tensor is a function of all six components of the strain tensor. Assuming the dependencies are linear leads to an equation with 36 constants. Thus, the generalised form of Hooke's law can be written as:

$$\sigma_{ik} = \sum_{nm} C_{iknm}\varepsilon_{nm} \tag{1}$$

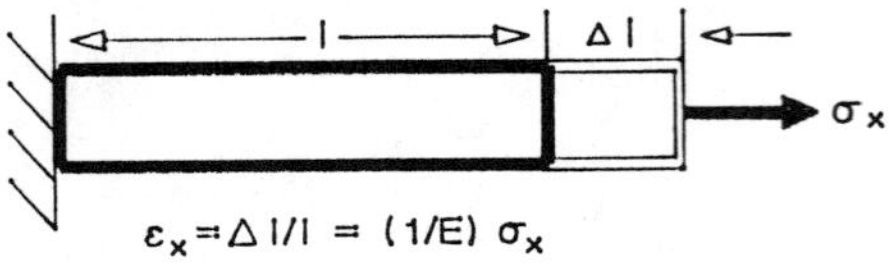

Fig. 1. Uniaxial elongation

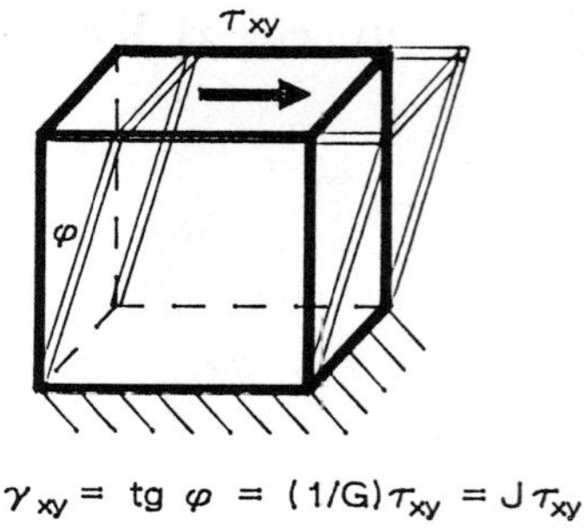

Fig. 2. Simple shear

The coefficient matrix of C_{iknm} determines a fourth-order tensor with 21 mutually independent coefficients. However, depending on the symmetry of the material under study, the complexity can be considerably reduced. For example, only 5 elastic constants are required to describe the linear stress/strain relationships for a hexagonal crystal; for a cubic crystal only 3 and for an isotropic solid only 2. In the case of uniaxial elongation or simple shear of an isotropic material Hooke's law has the following simple form (Figs. 1 and 2):

$$\sigma_x = E\varepsilon_x \tag{2}$$

$$\tau_{xy} = G\gamma_{xy} \tag{3}$$

With E: Modulus of elasticity (Young's modulus), G: Shear modulus, J: Shear compliance.

For an ideal liquid, by analogy, a proportionality exists between the stress and the strain rate. Under shear the relationship is given by Newton's law of friction. Thus, if of two parallel plates separated by a layer of liquid, the upper plate moves with a constant velocity v_x in the direction x, a shear gradient $\Delta v_x/\Delta y$ will develop due to adhesion of the liquid to the surface of the plate (Fig. 3).

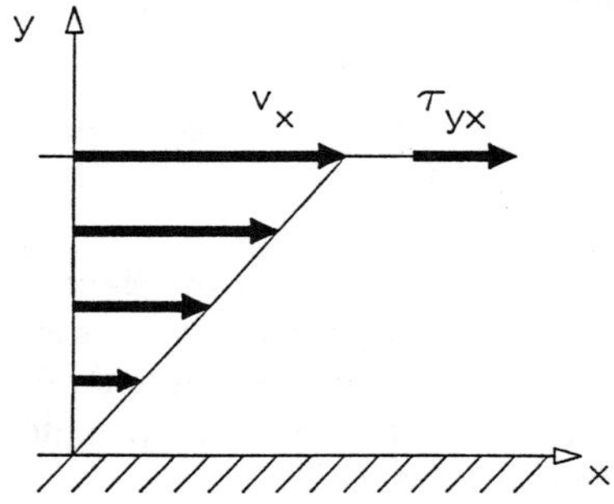

Fig. 3. Shear gradient in a Newtonian fluid

If $\tau_{yx} = \tau_{xy}$ is the shear stress operating on the moving plate, then:

$$\tau_{xy} \sim \frac{\Delta v_x}{\Delta y} \tag{4}$$

With $\Delta v_x/\Delta y = \Delta x/\Delta t \Delta y = \dot{\gamma}_{xy}$ the Newtonian law:

$$\tau_{xy} = \eta \dot{\gamma}_{xy} \tag{5}$$

results. In direct analogy Trouton's law for extension flow

$$\sigma_x = \mu \dot{\varepsilon}_x \tag{6}$$

is valid. Here $\dot{\gamma}_{xy}$: Shear gradient or shear rate, η: Shear viscosity, $\dot{\varepsilon}_x$: Extension velocity, μ: Extension viscosity.

For non-compressible solids or liquids the following relationships hold:

$$E = 3\,G \quad \text{and} \tag{7}$$

$$\mu = 3\,\eta \tag{8}$$

Table 1 is a tabulation of E-moduli for a selection of typical construction materials.

Table 1. E-Moduli of some typical construction materials (from [2, 3, 85])

Substance	E-Modul/GPa	Substance	E-Modul/GPa
Metals:		Glass-reinforced polymers:	
Aluminium	70.3	LDPE	1.8– 3.2
Copper	129.9	HDPE	3.2– 6.7
Iron	211.9	Polypropylene	3.2– 6.5
Titanium	115.7	Polycarbonate	7.0–13.5
Tungsten	411.0	Polyamide 6,6	5.0–13.5
		Polyester resins	10.0–35.0
Glass	80.1	ABS	5.0– 9.0
Silicon carbide	470.0		
Diamond	965.0	LC-Polymers	50–100
Quartz	73.1	Kevlar fibers	100–180
Polymers:		Elastomers:	
LDPE	0.2–0.5	Unfilled rubber	0.001–0.005
HDPE	0.7–1.4	Filled rubber	0.01 –0.05
Polypropylene (isotactic)	1.1–1.3		
Polycarbonate	2.1–2.4		
Polyamide 6,6	1.6–3.4		
PMMA	2.7–3.2		
Polystyrene	3.2–3.25		
ABS	1.5–2.7		

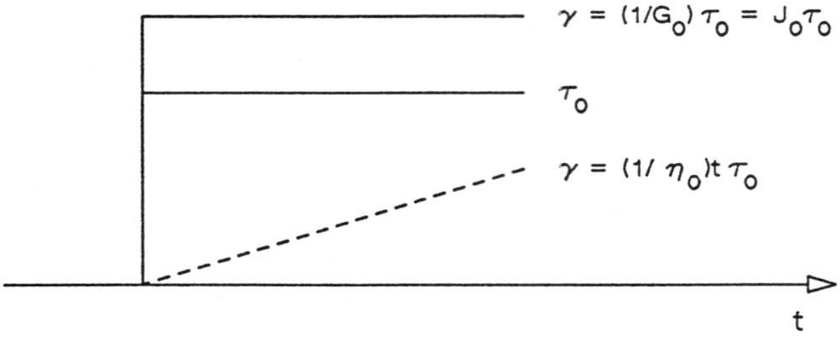

Fig. 4. Deformation behavior of an ideally elastic solid or an ideal liquid following an instantaneous increase in applied stress

For ideally elastic solids and ideal liquids (see Fig. 4) the moduli and viscosities are real material constants and these are independent of the course of the deformation.

Such ideal solids or liquids represent extremes; real solids undergo viscous flow and all liquids exhibit elasticity. For polymeric materials the two properties are superimposed in a complex fashion so that their behavior under deformation is called visco-elastic. Both phenomenological theories and those which are based on atomic models (e.g. the theory of the jump model) rely on Hooke's law except that, rather than real constants, the moduli are time or frequency dependent functions.

2.2 Relaxation and Creep Experiments on Polymers

In order to characterize an instantaneous application of a stress or strain, the quantity $e_0(t)$ as defined by:

$$e_0(t) = \begin{cases} 0 & \text{for } t < 0 \\ 1 & \text{for } t > 0 \end{cases} \tag{9}$$

can be used.

2.2.1 Creep Experiment

In a creep experiment an instantaneous stress is applied to the sample:

$$\sigma(t) = \sigma_0 e_0(t) \tag{10}$$

This stress-jump induces a time-dependent creep:

$$\varepsilon(t) = \frac{1}{M(t)} \sigma_0 = J(t)\sigma_0 \tag{11}$$

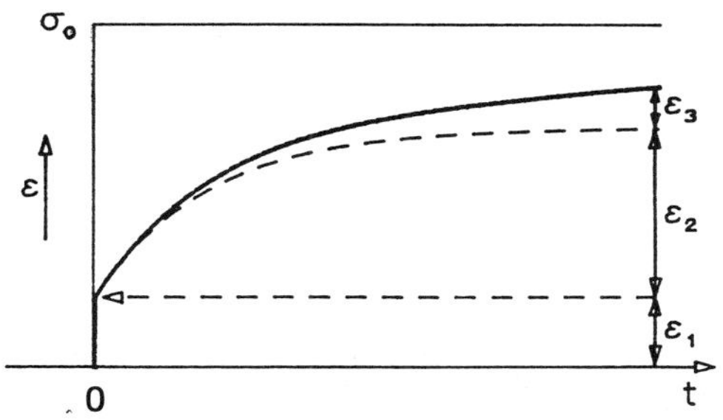

Fig. 5. Creep following a stress-jump

where M(t) is the function of the modulus appropriate for the stress applied and J(t) is the creep or retardation function which is also called the compliance.

The total creep comprises three elements of deformation (see Fig. 5):

$$\varepsilon_1 = J_0\sigma_0 \quad \text{Hookean deformation} \tag{12a}$$

$$\varepsilon_2 = \sum_k \Delta J_k(1 - e^{-t/\tau_{\varepsilon k}})\sigma_0 \quad \text{"Highly elastic" deformation (Relaxation component)} \tag{12b}$$

$$\varepsilon_3 = (t/\eta)\sigma_0 \quad \text{Newtonian flow} \tag{12c}$$

The sum in Eq. 12b takes account of the fact that real creep processes do not take place with a single time-constant: They are characterized by a spectrum of relaxation times. If a continuous relaxation-time spectrum is to be considered then the sum in Eq. 12b can be replaced by an integral:

$$J(t) = J_0 + \int_{-\infty}^{+\infty} L(\ln \tau)(1 - e^{-t/\tau})d \ln \tau + t/\eta_0 \tag{13}$$

2.2.2 Relaxation Experiment

For a relaxation experiment a sample is suddenly extended:

$$\varepsilon(t) = \varepsilon_0 e_0(t) \tag{14}$$

For the corresponding, time-dependent stress Hooke's law gives:

$$\sigma(t) = M(t)\varepsilon_0 \tag{15}$$

As can be seen from Fig. 6, the corresponding modulus function M(t) is given by:

$$M(t) = M_0\delta(t) + M_\infty + \sum_k \Delta M_k e^{-t/\tau_{\sigma k}} \tag{16}$$

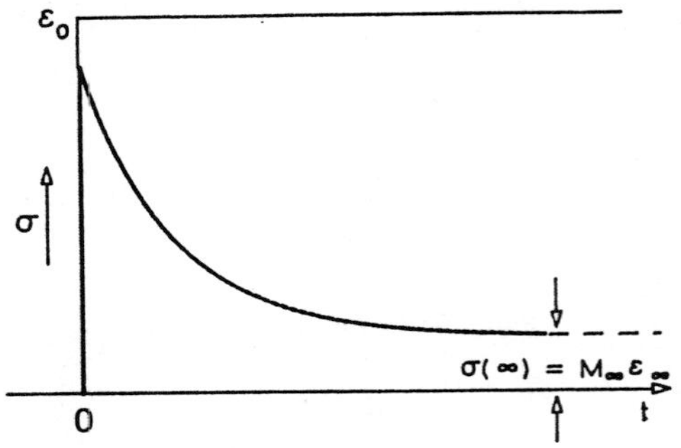

Fig. 6. Stress relaxation after a sudden deformation

or, if a continuous relaxation time spectrum is considered:

$$M(t) = \int_{-\infty}^{+\infty} H(\ln \tau) e^{-t/\tau} \, d \ln \tau + M_\infty \tag{17}$$

2.2.3 Basic Law for Relaxation and Creep

If the pure viscous flow processes are neglected then the time-dependent base functions retain only an elastic and a relaxation component. Thus, as $t \to \infty$, the stress and the deformation become proportional to one another:

$$\frac{\sigma_\infty}{\varepsilon_\infty} = M_\infty \tag{18}$$

In this case the stress-strain behavior can be described very simply by:

$$\dot{\varepsilon} = \frac{1}{\tau_\varepsilon}(\varepsilon_\infty - \varepsilon) \tag{19a}$$

$$\dot{\sigma} = \frac{1}{\tau_\sigma}(\sigma_\infty - \sigma) \tag{19b}$$

By eliminating the terms ε_∞ and σ_∞ from Eqs. 19a and 19b by substitution using Eq. 18, one obtains the fundamental equation for a "simple relaxing body":

$$\sigma + \tau_\sigma \dot{\sigma} = M_\infty(\varepsilon + \tau_\varepsilon \dot{\varepsilon}) \tag{20}$$

2.3 Dynamic Relaxation Experiments [4, 5]

If a sinusoidal load is applied to a sample the force and the deformation of the sample are out of phase by an amount δ:

$$\sigma(t) = \sigma_0 e^{i\omega t} \tag{21a}$$

$$\varepsilon(t) = \varepsilon_0 e^{i(\omega t - \delta)} \tag{21b}$$

This out-of-phase phenomenon can be described by a complex modulus $M^* = M' + iM''$:

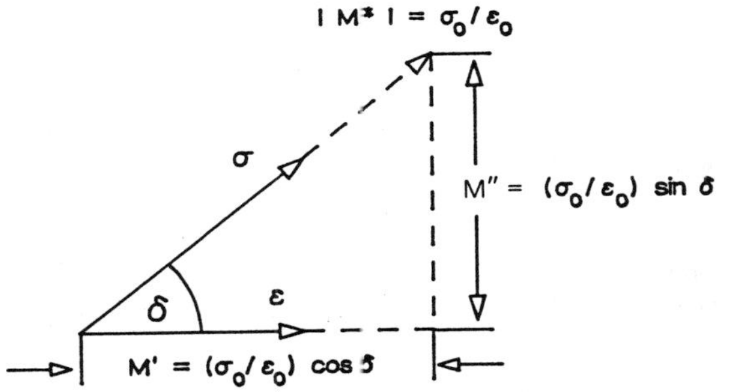

Fig. 7. Vector diagram of the complex modulus

Substitution of Eqs. 21a and 21b into Eq. 20 with $M_{t\to\infty} \triangleq M_{\omega\to 0}$ gives:

$$\sigma(1 + i\omega\tau_\sigma) = M_0\varepsilon(1 + i\omega\tau_\varepsilon) \tag{22}$$

$$M_\infty = \frac{\sigma(\omega \to \infty)}{\varepsilon(\omega \to \infty)} = M_0 \frac{\tau_\varepsilon}{\tau_\sigma} \tag{23}$$

Since $M_\infty > M_0$, it follows that τ_ε is always larger than τ_σ.

The difference $\Delta M = M_\infty - M_0$ is called the relaxation strength. At any frequency Eq. 22 can be used to obtain the complex, dynamic modulus:

$$M^*(\omega) = \frac{\sigma(\omega)}{\varepsilon(\omega)} = M'(\omega) + iM''(\omega) \tag{24}$$

with

$$M'(\omega) = M_0 + \Delta M \frac{\omega^2\tau^2}{1 + \omega^2\tau^2} \tag{25a}$$

$$M''(\omega) = \Delta M \frac{\omega\tau}{1 + \omega^2\tau^2} \tag{25b}$$

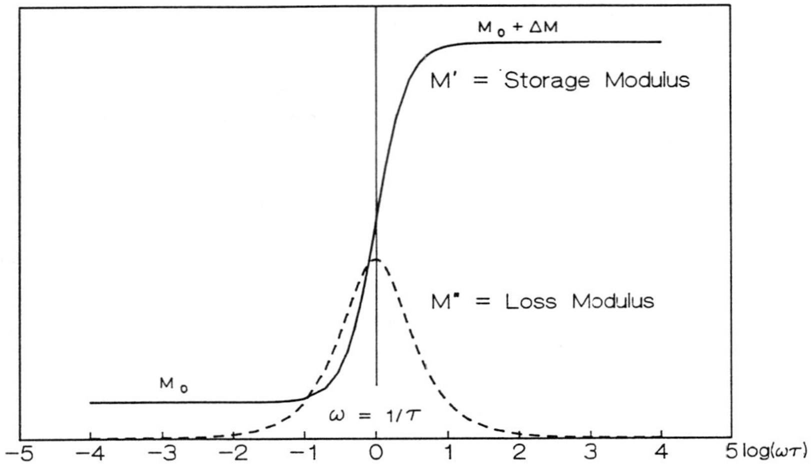

Fig. 8. The components of the complex modulus $M^* = M' + iM''$ as a function of frequency for a simple relaxation process (Eqs. 25a and b)

The ratio $M''/M' = \tan\delta$ is called the loss factor. Figure 8 shows the components of the complex modulus as a function of the frequency (Eqs. 25a and b) for a simple relaxation process.

2.4 Technical Measures for Damping [6]

2.4.1 Energy Dissipation Under Defined Load Conditions

With every cycle some of the applied energy ΔW is converted, irreversibly, into heat. This loss energy can be represented by the area enclosed by an ellipse produced when σ and ε are plotted in a right-angled coordinate system:

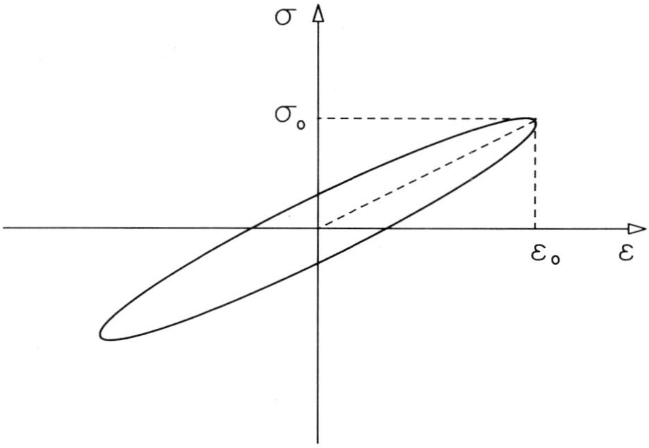

Fig. 9. Energy loss during a single period of dynamic stress

The area enclosed by such an ellipse is given by:

$$\Delta W = \oint \sigma d\varepsilon = \pi \sigma_0 \varepsilon_0 \sin \delta \tag{26}$$

Depending on the nature of the applied load different equations can be derived for the dissipated energy:

a) **At constant deformation:** $\varepsilon_{01} = \varepsilon_{02} = \varepsilon_{03} = \ldots$

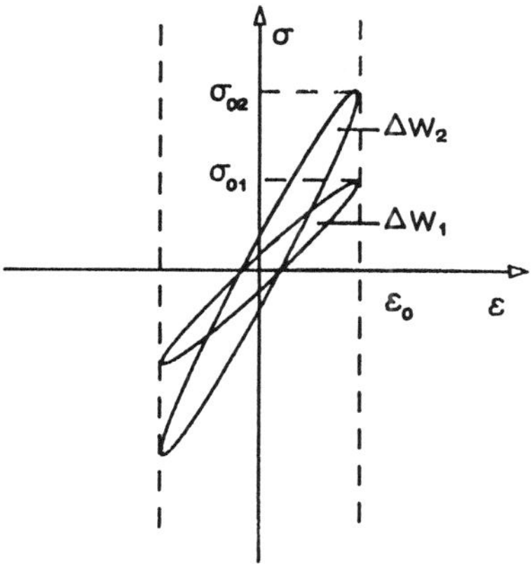

Fig. 10. Energy dissipated at constant deformation

$$\Delta W = \pi \varepsilon_0^2 M_0 \sin \delta = \pi M'' \varepsilon_0^2 \tag{27}$$

Thus, for dynamic loading with constant deformation the loss modulus is the decisive quantity:

$$\Delta W_1 / \Delta W_2 = M_1'' / M_2'' \tag{28}$$

b) **At constant stress:** $\sigma_{01} = \sigma_{02} = \sigma_{03} = \ldots$

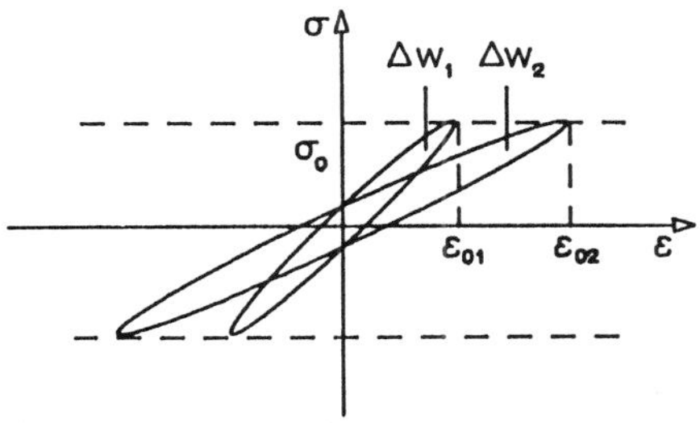

Fig. 11. Energy dissipated at constant stress

$$\Delta W = \pi \sigma_0^2 \frac{\sin \delta}{M_0} = \pi J'' \sigma_0^2 \tag{29}$$

The energy dissipated during dynamic loading with a constant stress is determined by the loss compliance:

$$\Delta W_1/\Delta W_2 = J_1''/J_2'' \quad (30)$$

c) **At** constant energy: $(\varepsilon_0\sigma_0)_1 = (\varepsilon_0\sigma_0)_2 = \ldots$

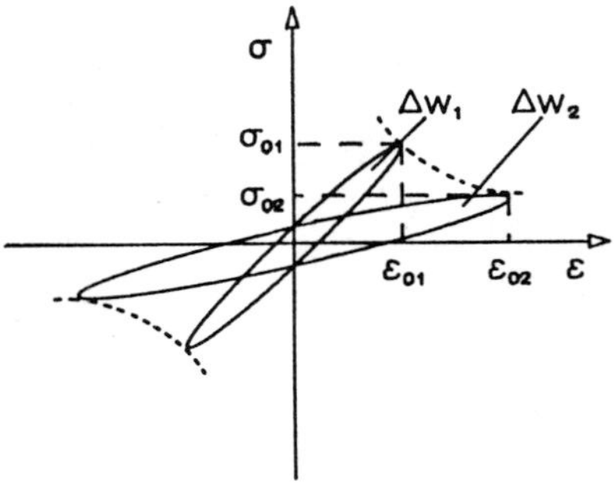

Fig. 12. Energy dissipated at constant energy input

$$\Delta W = \pi\sigma_0\varepsilon_0 \sin \delta \quad (31)$$

During dynamic loading with the applied energy held constant the sin δ (or, if the phase angle is small, the loss factor tan δ) will determine the loss energy:

$$\Delta W_1/\Delta W_2 = \sin \delta_1/\sin \delta_2 \cong \tan \delta_1/\tan \delta_2 \quad (32)$$

The rolling resistance of vehicle tires is due primarily to processes involving energy dissipation from the polymer network of the tread. Since the rolling resistance has a

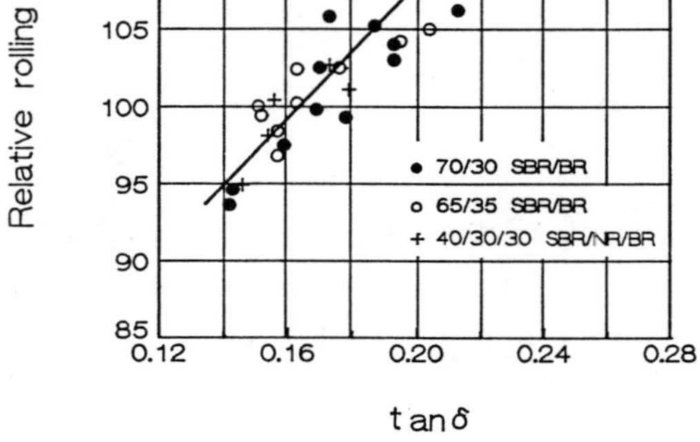

Fig. 13. Rolling resistance as a function of tan δ
Strain amplitude: 7,5% ptp
Frequency: 10 Hz
Temperature: 50°C
(From [7])

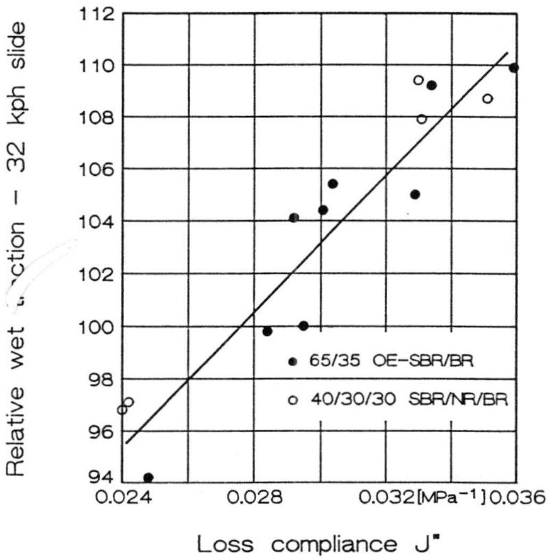

Fig. 14. Wet traction as a function of loss compliance
Strain amplitude: 25% ptp
Frequency: 1 Hz
Temperature: 0°C
(From [7])

direct influence on the amount of fuel consumed, there is considerable interest in minimizing such hysteresis losses. On the other hand, the same processes result in an improved wet traction so that the development of the optimum tread compounds inherently involves a fine compromise. As can be seen from Figs. 13 and 14, there is a linear relationship between both the rolling resistance and the wet traction of a tire and the loss factor or the loss compliance respectively. Thus, by making dynamic-mechanical measurements on model vulcanisates the heavy costs involved in making and testing complete tires can be considerably reduced.

2.4.2 Rebound Elasticity

Rebound elasticity, as determined with the aid of an oscillating pendulum, is a particularly useful measure for the damping quality of materials employed by the rubber industry. From the difference ΔW in the kinetic energy of the pendulum before (W_0) and after (W_1) hitting the sample, the rebound elasticity R is given by:

$$R = 1 - \frac{\Delta W}{W_0} \tag{33}$$

Since the test involves a free, damped oscillation:

$$\begin{aligned} R &= h_1/h_0 = h_2/h_1 = h_3/h_2 = \ldots \\ &= W_1/W_0 = W_2/W_1 = W_3/W_2 = \ldots \end{aligned} \tag{34}$$

The amplitude of such oscillations decreases exponentially so that:

$$R = \frac{h_{n+1}}{h_n} = 1 - \frac{\Delta W}{W_0} = e^{-\Lambda} \quad \text{or} \tag{35}$$

$$\Lambda = \ln \frac{1}{1 - \frac{\Delta W}{W_0}} \approx \ln\left(1 + \frac{\Delta W}{W_0}\right) \tag{36}$$

Expanding Eq. 36 as a series leads to:

$$\Lambda = \frac{\Delta W}{W_0} - \frac{1}{2}\left(\frac{\Delta W}{W_0}\right)^2 + \frac{1}{3}\left(\frac{\Delta W}{W_0}\right)^3 - + \ldots \tag{37}$$

With $\Lambda = \pi \tan \delta$, a first approximation ($\Delta W/W_0 \ll 1$) can be written as:

$$R = 1 - \pi \tan \delta \tag{38}$$

3 Simple Phenomenological Models [4, 8]

Many attempts have been made to describe the creep and relaxation behaviour of polymers in terms of simple models based on elastic springs and viscous damping elements.

3.1 Maxwell's Model

Maxwell's model, in its simplest form, is a single spring attached in series to a damping cylinder:

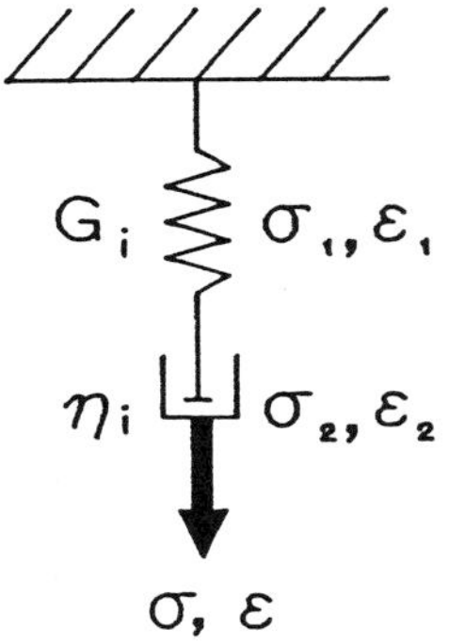

Fig. 15. Maxwell element

The premises for this model give rise to the two conditions:

$$\sigma = \sigma_1 = \sigma_2 \quad \text{and} \tag{39}$$

$$\varepsilon = \varepsilon_1 + \varepsilon_2 \tag{40}$$

After differentiating Eq. 40 with respect to time and using the two basic equations of mechanics (Eqs. 2 and 5) one obtains:

$$\dot{\varepsilon} = \frac{\dot{\sigma}}{G_i} + \frac{\sigma}{\eta_i} \tag{41}$$

Setting the stress equal to a periodical function, i.e. $\sigma = \sigma_0 \exp i\omega t$, one obtains:

$$\dot{\varepsilon} = \frac{i\omega\sigma}{G_i} + \frac{\sigma}{\eta_i} \quad \text{or} \tag{42}$$

$$i\omega\varepsilon = \left(\frac{i\omega}{G_i} + \frac{1}{\eta_i}\right)\sigma \tag{43}$$

The real and imaginary components of the complex modulus can now be obtained with the aid of Eq. 25:

$$G^*(\omega) = \frac{i\omega\eta_i}{1 + i\omega\tau_i} \quad \text{with} \quad \tau_i = \eta_i/G_i \tag{44}$$

$$G'(\omega) = G_i \frac{\omega^2\tau_i^2}{1 + \omega^2\tau_i^2} \tag{44a}$$

$$G''(\omega) = G_i \frac{\omega\tau_i}{1 + \omega^2\tau_i^2} \tag{44b}$$

$$\tan\delta = 1/\omega\tau_i \tag{44c}$$

3.2 Kelvin-Voigt Model

The Kelvin-Voigt model differs from Maxwell's in that the damping cylinder and the spring are connected in parallel rather than in series:

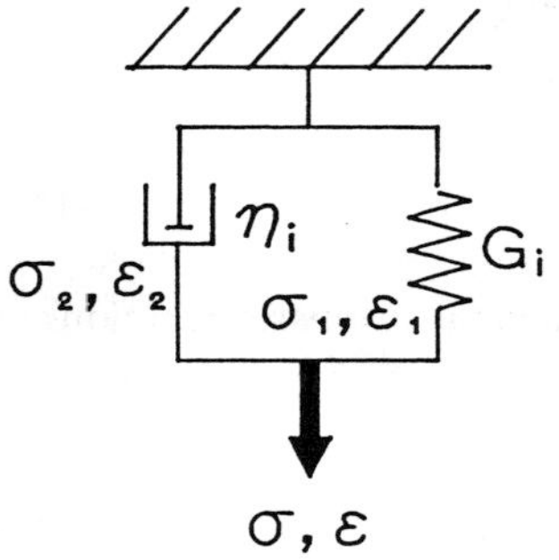

Fig. 16. Kelvin-Voigt element

For this model

$$\varepsilon = \varepsilon_1 = \varepsilon_2 \quad \text{and} \tag{45}$$

$$\sigma = \sigma_1 + \sigma_2 \tag{46}$$

$$\sigma = \eta_i \dot{\varepsilon} + G_i \varepsilon \tag{47}$$

In the same way as for Maxwell's model, using $\sigma = \sigma_0 \exp i\omega t$ leads to:

$$\sigma = (i\omega\eta_i + G_i)\varepsilon \tag{48}$$

from which the components of the complex modulus

$$G'(\omega) = G_i \tag{49a}$$

$$G''(\omega) = \omega\eta_i \tag{49b}$$

$$\tan \delta = \omega\tau_i \tag{49c}$$

can be derived.

Furthermore, the following relationship exists, between the complex modulus and the complex compliance:

$$J^* = \frac{1}{G^*} \tag{50}$$

Thus, as a consequence of the Kelvin-Voigt approach, the components of the compliance are given by:

$$J' = \frac{J_i}{1 + \omega^2\tau_i^2} \tag{50a}$$

$$J'' = J_i \frac{\omega\tau_i}{1 + \omega^2\tau_i^2} \tag{50b}$$

In the following table (Table 2) the various equations resulting from these two models are summarized for easy comparison.

These two models are only of limited use in describing the behavior of real systems. Thus, Maxwell's model cannot account for the time-dependent aspect of creep and that of Kelvin-Voigt fails to account for stress relaxation.

Table 2. Viscoelastic Functions of Modulus and Compliance

Maxwell	Kelvin-Voigt
$G(t) = G_i e^{-t/\tau_i}$	$G(t) = G_i$
$J(t) = J_i + t/\eta_i$	$J(t) = J_i(1 - e^{-t/\tau_i})$
$G'(\omega) = G_i \dfrac{\omega^2\tau_i^2}{1 + \omega^2\tau_i^2}$	$G'(\omega) = G_i$
$G''(\omega) = G_i \dfrac{\omega\tau_i}{1 + \omega^2\tau_i^2}$	$G''(\omega) = \omega\eta_i = G_i\omega\tau_i$
$\eta'(\omega) = \dfrac{\eta_i}{1 + \omega^2\tau_i^2}$	$\eta'(\omega) = \eta_i$
$J'(\omega) = J_i = 1/G_i$	$J'(\omega) = \dfrac{J_i}{1 + \omega^2\tau_i^2}$
$J''(\omega) = 1/\omega\eta_i$	$J''(\omega) = J_i \dfrac{\omega\tau_i}{1 + \omega^2\tau_i^2}$
$\tan\delta = 1/\omega\tau_i$	$\tan\delta = \omega\tau_i$

3.3 Relaxation and Retardation Spectra

The course of any relaxation or creep process in real polymers involves not a single, but rather a broad spectrum of relaxation and retardation times.

If a number n of Maxwellian elements, each having a discrete relaxation time τ_i, and a discrete relaxation strength G_i are connected in parallel (see Fig. 17), the resulting force is the simple sum of the forces required to deform each individual element. Therefore, $G(t)$, $G'(\omega)$, $G''(\omega)$ and $\eta'(\omega)$ can be obtained by simple summation of the corresponding terms for the individual elements.

$$G(t) = \sum_{i=1}^{n} G_i\, e^{-t/\tau_i} \tag{51}$$

Alternatively, a number of n Kelvin-Voigt elements, each with a discrete retardation time τ_i and compliance J_i can be connected in series (see Fig. 18) so that the total elongation is given by the sum of the elongations of the individual elements. Consequently, $J(t)$, $J'(\omega)$ and $J''(\omega)$ can be obtained by simple summation of the terms for the individual elements:

$$J(t) = \sum_{i=1}^{n} J_i(1 - e^{-t/\tau_i}) \tag{52}$$

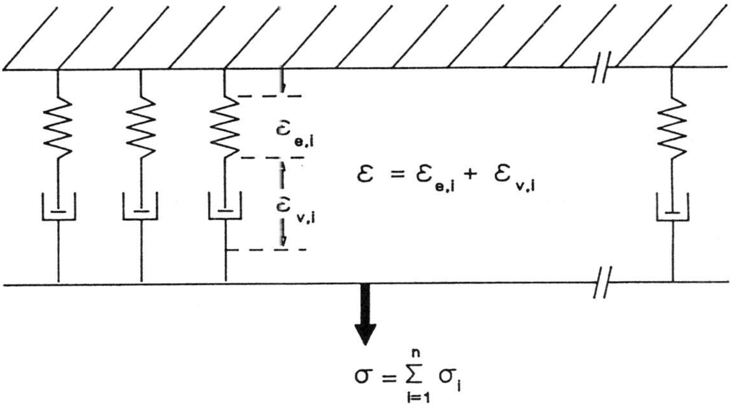

Fig. 17. Generalized Maxwell model

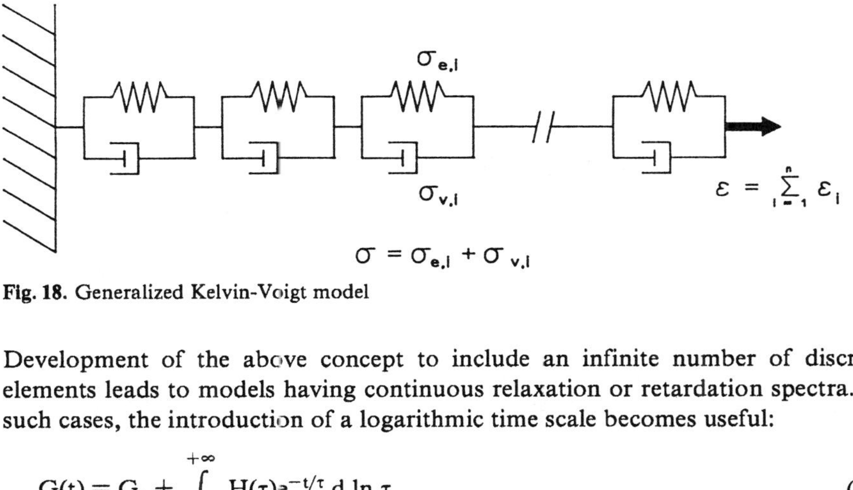

Fig. 18. Generalized Kelvin-Voigt model

Development of the above concept to include an infinite number of discrete elements leads to models having continuous relaxation or retardation spectra. In such cases, the introduction of a logarithmic time scale becomes useful:

$$G(t) = G_e + \int_{-\infty}^{+\infty} H(\tau) e^{-t/\tau} \, d \ln \tau \tag{53}$$

The equilibrium modulus G_e of a visco-elastic solid accounts for a discrete contribution to the time spectrum where $\tau = \infty$. For an ideal liquid $G_e = 0$.

$$J(t) = J_g + \int_{-\infty}^{+\infty} L(\tau)(1 - e^{-t/\tau}) d \ln \tau + t/\eta_0 \tag{54}$$

The instantaneous compliance J_g corresponds to a discrete element of the spectrum with $\tau = 0$. Under dynamic load the following relationships can be obtained:

$$G' = G_e + \int_{-\infty}^{+\infty} H(\tau) \omega^2 \tau^2 / (1 + \omega^2 \tau^2) d \ln \tau; \tag{55}$$

$$G'' = \int_{-\infty}^{+\infty} H(\tau)\omega\tau/(1 + \omega^2\tau^2)d \ln \tau; \tag{56}$$

$$\eta' = \int_{-\infty}^{+\infty} H(\tau)\tau/(1 + \omega^2\tau^2)d \ln \tau; \tag{57}$$

$$J' = J_g + \int_{-\infty}^{+\infty} L(\tau)/(1 + \omega^2\tau^2)d \ln \tau; \tag{58}$$

$$J'' = \int_{-\infty}^{+\infty} L(\tau)\omega\tau/(1 + \omega^2\tau^2)d \ln \tau + 1/\omega\eta_0. \tag{59}$$

The steady-flow viscosity is defined as:

$$\eta_0 = \lim_{\omega \to 0} \left(\frac{G''}{\omega}\right) = \int_{-\infty}^{+\infty} H(\tau)\tau d \ln \tau \tag{60}$$

Furthermore, the instantaneous modulus $G_g(\omega \to \infty)$ is given by:

$$G_g = G_e + \int_{-\infty}^{+\infty} H(\tau)d \ln \tau \tag{61}$$

and the equilibrium or steady state compliance $J_e^0(\omega \to 0)$ can be written as:

$$J_e^0 = J_g + \int_{-\infty}^{+\infty} L(\tau)d \ln \tau \tag{62}$$

3.4 Approximate Determination of Relaxation Spectra [9, 10]

3.4.1 Method of Schwarzl and Stavermann [9]

Schwarzl and Stavermann start by taking a modulus function corresponding to a continuous relaxation time spectrum:

$$\begin{aligned} G(t) &= G_e + \int_{-\infty}^{+\infty} H(\tau)e^{-t/\tau}d \ln \tau \\ &= G_e + \int_{0}^{+\infty} h(\tau)e^{-t/\tau}d\tau \end{aligned} \tag{63}$$

whereby: $H(\tau) = \tau h(\tau)$ and $d \ln \tau = (1/\tau)d\tau$.

Equation 63 is a Laplace transformation from which, by inversion, $h(\tau)$ could be calculated. However, since the modulus curves $G(t)$ cannot be exactly determined analytically, inversion of the Laplace transformation can only be accomplished via appropriate approximations. Thus, initially, the usual form of the Laplace transformation is introduced:

$$f(t) = G(t) - G_e = \int_0^{+\infty} F(s)e^{-st}ds \tag{64}$$

with $\tau = s^{-1}$; $d\tau = -s^{-2}ds$

and $F(s) = s^{-2}h(\tau) = \tau^2 h(\tau)$

For the inverse of the Laplace transformation the following approximations can be introduced:

$$F(s) = \lim_{k\to\infty} \frac{(-1)^k}{k!}\left(\frac{k}{s}\right)^{k+1} f^{(k)}\left(\frac{k}{s}\right) \tag{65}$$

First approximation (k = 1): $F_1(s) = -\frac{1}{s^2} f'\left(\frac{1}{s}\right)$ (65a)

Second approximation (k = 2): $F_2(s) = \frac{4}{s^3} f''\left(\frac{2}{s}\right)$ (65b)

Third approximation (k = 3): $F_3(s) = -\frac{27}{2s^4} f'''\left(\frac{3}{s}\right)$ (65c)

An initial value for $h_k(\tau)$ can now be obtained:

$$1.\ h_1(\tau) = s^2F_1(s) = -f'\left(\frac{1}{s}\right) = -\left[\frac{dG(t)}{dt}\right]_{t=\tau} \tag{66}$$

$$2.\ h_2(\tau) = 4t\, f''(2t) = \left[4t\,\frac{d^2G(2t)}{d(2t)^2}\right]_{t=\tau} \tag{67}$$

From Eqs. 66 and 67 and using a logarithmic time axis and the appropriate differential operators one obtains:

$$\frac{d}{dt} = \frac{1}{t}\frac{d}{d\ln t} \quad \text{and} \quad \frac{d^2}{dt^2} = -\frac{1}{t^2}\frac{d}{d\ln t} + \frac{1}{t^2}\frac{d^2}{d(\ln t)^2}$$

$$H_1(\tau) = -\left[t\,\frac{dG(t)}{dt}\right]_{t=\tau} = -\left[\frac{dG(t)}{d\ln t}\right]_{t=\tau} \tag{68}$$

$$H_2(\tau) = \left[- \frac{dG(2t)}{d \ln t} + \frac{d^2G(2t)}{d (\ln t)^2} \right]_{t=\tau} \tag{69}$$

3.4.2 Method According to Ferry and Williams [10]

Employing a jump-function for $e^{-t/\tau}$ in the modulus function (Eq. 63)

$$e^{-t/\tau} = \begin{cases} 0 & \text{for } t < \tau \\ 1 & \text{for } t > \tau \end{cases} \tag{70}$$

With Eq. 70 the integral of Eq. 63 can now be written:

$$G(t) = G_e + \int_{\ln t}^{\infty} H(\tau) d \ln \tau \tag{71}$$

If $H(\tau) = 0$ for $\tau \to \infty$, then an initial approximate solution for Eq. 71 can be obtained:

$$H_1(\tau) = - \left[\frac{dG(t)}{d \ln t} \right]_{t=\tau} \tag{72}$$

Comparison of Eq. 72 with Eq. 68 shows that the first approximation obtained in this way is the same as that obtained by Schwarzl and Stavermann.

A second approximation assumes that $H_1(\tau)$, at least within narrow limits, can be set equal to the exponential function:

$$H_1(\tau) = kt^{-m} \tag{73}$$

If this function is substituted into Eq. 63 one obtains:

$$G(t) = G_e + kt^{-m}\Gamma(m) \tag{74}$$

where $\Gamma(m)$ is the gamma function:

$$\Gamma(s+1) = \int_0^{\infty} x^s e^{-x} dx \tag{75}$$

Thus, according to Ferry, the second approximation yields:

$$H_2(\tau) = \left[\frac{H_1(\tau)}{\Gamma(m+1)} \right]_{t=\tau} \tag{76}$$

Equation 76 prescribes a method for obtaining the relaxation time spectrum: For $\Gamma = 1$ one obtains a first approximation for $H_1(\tau)$. From a double logarithmic plot of $H_1(\tau)$ the slope $d \lg H_1(\tau)/d \lg \tau$ is equal to $-m$ (cf. Eq. 73) and enables a calculation of the correction term $\Gamma(m+1)$ in Eq. 76. An example of a relaxation time spectrum of the glass transition of PVC according to this approach is shown in Fig. 19.

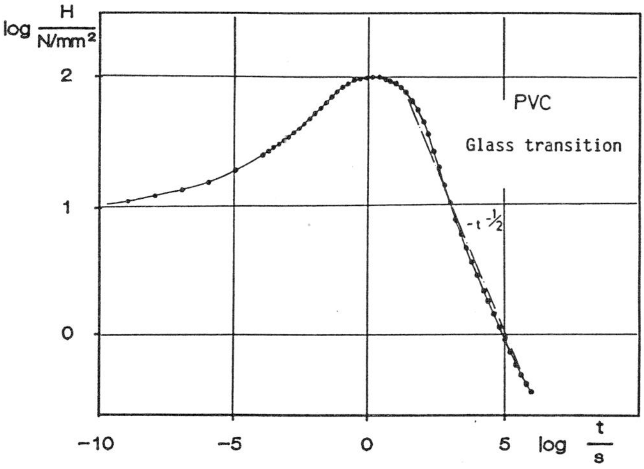

Fig. 19. Relaxation time spectrum of the glass transition of PVC

4 Molecular Models of Relaxation Behavior

4.1 Simple Jump Model

Polymer relaxation phenomena result from thermally activated jump processes involving individual molecules or molecular segments. A useful model for a simple change of site process is the Snoek effect observed in C-doped α-iron. The C-atoms are foreign elements in the cubic space-centered Fe-lattice and occupy sites half-way along the edges of the elementary cell (see Fig. 20).

In a cubic lattice there are three energetically identical positions with respect to the crystallographic axes (1, 2 and 3). In the absence of an external stress the foreign atoms will become randomly distributed between these positions. In contrast, should an external mechanical stress be applied, for example along the 1-axis, then the positions in this direction will be sites of potential minima and as such preferred sites for the foreign atoms. These will then "jump" appropriately as exemplified by the arrow in Fig. 20. Should the direction of the stress be reversed, the foreign atoms will return to their initial positions. If the frequency of the periodically changing stress field is low, all the C-atoms will have time to change their position within one half of the oscillation period. If all the C-atoms are in preferred positions an additional macroscopic deformation will occur and a correspondingly larger compliance results. If, on the other hand, the frequency of the stress field is too high to allow the C-atoms to reach the preferred positions a reduced macroscopic deformation is observed: the compliance decreases. Thus, with increasing frequency a stepwise increase in modulus occurs at a frequency corresponding to the jump rate (see Fig. 8).

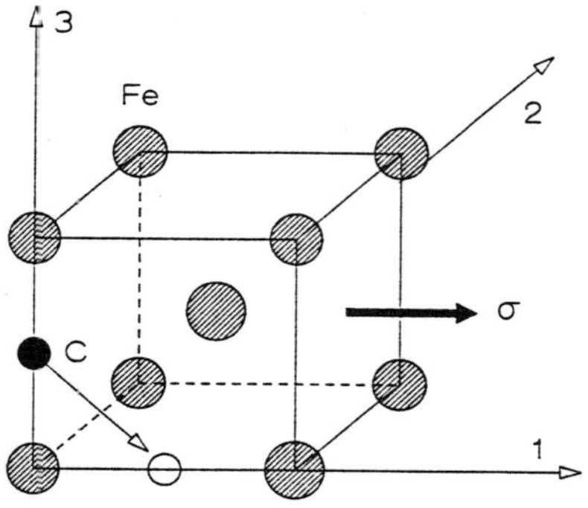

Fig. 20. Snoek effect in C-doped α-iron

4.2 Change of Position in Terms of a Potential Model

On a microscopic level this model considers the potential energy associated with two equilibrium lattice sites for a C-atom capable of changing its position. A simplified, one- dimensional model is shown in Fig. 21.

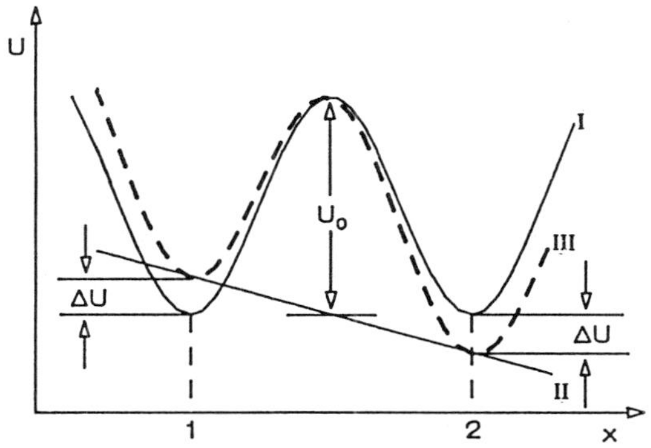

Fig. 21. Change of position in a one dimensional potential model.
I Potential in the absence of an external stress field, II Potential of the external stress field, III Total potential

Initially, the C-atom oscillates in a potential trough with a frequency ν_0 and an energy of the order of magnitude $kT \ll U_0$. As predicted by the Boltzmann distribution, this atom will, at some time, attain an energy large enough to enable it to overcome the potential barrier; i.e., it will become thermally activated. In the same fashion, the atom can return to its initial position. In the absence of an external field the probability that an atom is in position 1 or 2 is equal i.e.,

$$N_1^0 = N_2^0 = N/2; \text{ N: Total number/cm}^3 \tag{77}$$

Thus, the jump rate (number of jumps per second) in the absence of an external field can be obtained with the aid of classical statistics*:

$$\Gamma^0 = g\nu_0 \exp(-U_0/kT) \tag{78}$$

where g is the number of statistically identical paths by which the atom can leave a trough ($g = 1$ in this example) and U_0 is the activation energy.

---

* Formally,

$$\Gamma^0 = g\nu_0 \exp(-\Delta G/kT) = g\nu_0 \exp(\Delta S/k) \exp(-\Delta H/kT)$$

so that an entropy term is also included in the frequency factor.

As the result of an applied field the frequency of the jumps from position 1 to 2 and from 2 to 1 will be different:

$$\Gamma_{\substack{12\\21}} = \nu_0 \exp(-U_0 \pm \Delta U)/kT \tag{79}$$

Expanding Eq. 79 and making a linear approximation leads to

$$\Gamma_{\substack{12\\21}} = \Gamma^0 (1 \pm \Delta U/kT) \tag{80}$$

The equations for the number of particles in positions 1 and 2 with respect to time

$$\dot{N}_1 = -N_1\Gamma_{12} + N_2\Gamma_{21} \tag{81}$$

$$\dot{N}_2 = -N_2\Gamma_{21} + N_1\Gamma_{12} \tag{82}$$

can then be obtained. Combining these equations leads to:

$$\frac{1}{2\Gamma^0}\frac{d}{dt}(N_2 - N_1) = -(N_2 - N_1) + \frac{\Delta U}{kT}(N_1 + N_2) \tag{83}$$

$$(N_1(0) = N_2(0) = N/2)$$

In analogy to a phenomenological approach, the solution of this differential equation gives rise to two cases:

a) A constant stress field is applied at time $t=0$:

$$N_2 - N_1 = N\frac{\Delta U}{kT}(1 - e^{-t/\tau}) \tag{84}$$

b) The applied field changes periodically ($\sim \exp i\omega t$):

$$N_2 - N_1 = N\frac{\Delta U}{kT}\frac{1}{1 + i\omega\tau} \tag{85}$$

with

$$\tau = \frac{1}{2\Gamma^0} = \frac{1}{2\nu_0}\exp(U_0/kT) \tag{86}$$

Equation 86 is the socalled Arrhenius relationship and enables the relaxation time, defined phenomenologically as the inverse of the jump rate, to be appreciated at a molecular level. The simple jump model also yields characteristic time functions (Eqs. 84 and 85), which are identical with those obtained phenomenologically (Eqs. 12b and 25).

4.3 Viscosity in Terms of the Simple Jump Model

The jump model describes the temperature dependence of the viscosity of low molar mass liquids and can, with the aid of suitable molecular model assumptions, be used to describe polymer melts. From the difference between the jump rates of individual molecules with respect to the direction of the external field (Eq. 80) the resulting relative displacement between two adjacent layers of the liquid can be calculated. This leads directly to an expression for the shear rate in a liquid. Thus, consider two layers of a liquid, one above the other, where the upper layer includes a void (Fig. 22).

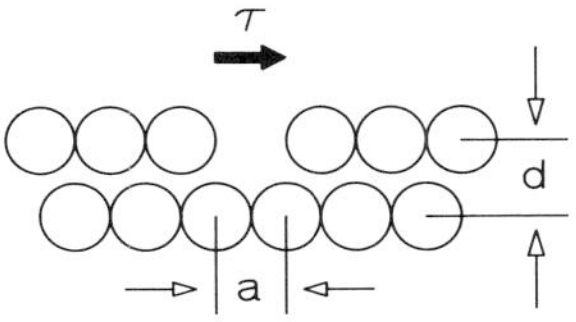

Fig. 22. Two liquid layers with a void

If a shear stress τ is operating in the upper layer, the molecules of this layer will move in consecutive jumps in the direction of the external field, which effects a sliding of the upper layer with respect to the lower one. The macroscopic shear deformation γ can be determined in analogy to the metallic dislocation theory. If N dislocations with a strength b occur per unit volume and cover an area F of their glide plane, a macroscopic shear deformation γ takes place:

$$\gamma = NFb = Nv \tag{87}$$

and

$$\dot{\gamma} = \dot{N}v \tag{88}$$

For a liquid, b corresponds to the distance between two equilibrium positions (a in Fig. 22) and v is an approximation of the volume of a molecule v_m. $\dot{N}$ is obtained from the difference between the jump rates:

$$\dot{N} = (\Gamma_{12} - \Gamma_{21})N \tag{89}$$

With Eq. 80 one can then obtain:

$$\dot{\gamma} = p_v \frac{2\Delta U}{kT} v_0 \exp(-U_0/kT) N v_m \tag{90}$$

Equation 90 has been expanded by the factor p_v, the probability that the potential trough adjacent to a thermally activated molecule is vacant. The oscillation frequency v_0 of a molecule with a mass m can be obtained directly from the periodic approximation for the potential energy U(x) (see also Fig. 21):

$$v_0 = \left(\frac{U_0}{2m}\right)^{1/2} \frac{1}{a} \tag{91}$$

The contribution of the applied field to the potential U is given by:

$$\Delta U = \frac{\tau v}{2} \tag{92}$$

With $Nv_m = 1$ and substituting Eqs. 90 and 91 into Eq. 89 the viscosity can be formulated as:

$$\eta = \tau/\dot{\gamma} = \frac{1}{p_v}\left(\frac{2m}{U_0}\right)^{1/2} \frac{kT}{v^{2/3}} \exp\left(U_0/kT\right) \tag{93}$$

In an approximation of the "free volume"-theory (see Sect. 5.2) the probability p_v is included arbitrarily with the term $\exp(-v^*/v_f)$ where v^* corresponds to the smallest volume necessary for a single diffusion jump and v_f is the average free volume per molecule. Thus, Eq. 93 can be rewritten:

$$\eta = v_0 \exp\left(v^*/v_f\right) \exp\left(U_0/kT\right) \tag{94}$$

Since v^* can be put equal to the volume of a molecule v_m for a first approximation, one obtains from Eq. 94, in terms of molar quantities:

$$\eta = \eta_0 \exp\left[V_m/(V - V_m) + Q/RT\right] \tag{95}$$

Here, the molar free volume V_f has been taken as being equal to the difference between the total volume V and the molecular volume V_m. Q is the energy of activation for a diffusion jump per mole.

4.4 Determining the Energy of Activation by Experiment

The Arrhenius equation (Eq. 86) gives the relationship between either time or frequency and temperature from which the activation energy for a relaxation process can be determined. During a dynamic experiment the maximum in the loss modulus curve occurs at:

$$\omega_{max}\tau = 1 \tag{96}$$

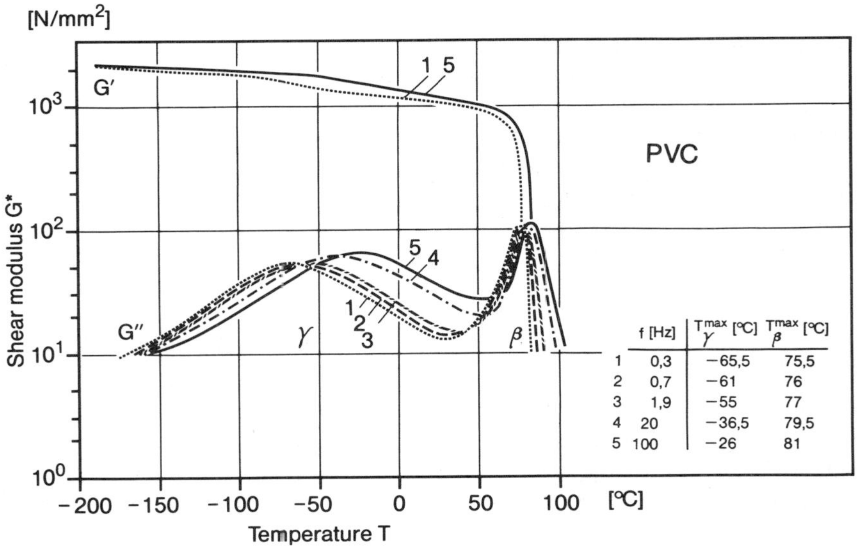

	f [Hz]	T_{γ}^{max} [°C]	T_{β}^{max} [°C]
1	0,3	−65,5	75,5
2	0,7	−61	76
3	1,9	−55	77
4	20	−36,5	79,5
5	100	−26	81

Fig. 23. Complex shear modulus G' and G'' of PVC as a function of temperature at different frequencies

and from Eq. 86 it follows that:

$$\omega_{max} = 2\nu_0 \exp(-U_0/kT_{max}) \tag{97}$$

or

$$\log f_{max} = -(Q/RT_{max}) \log e + \log f_0 \tag{98}$$

Thus, with increasing frequency the damping maximum of a relaxation process is shifted to higher temperatures. Figure 23 shows such a shift for PVC. The preexponential factor $f_0 = \nu_0/\pi$ can, as a first approximation, be assumed to be constant; for polymeric materials it lies between 10^{12} and 10^{14}. Both low temperature processes (γ-process) and glass transitions (β-process or main dispersion) are shifted with increasing frequency to higher temperatures. If the logarithm of the frequencies of the loss maxima are plotted against the reciprocal temperatures of these maxima, a straight line is obtained, the gradient of which is directly proportional to the activation energy Q (in Eq. 98 for 1 Mole) of the particular relaxation process. Figure 24 shows an Arrhenius plot for PVC where, in addition to the mechanical data, the data obtained from dielectric relaxation measurements (Im (1/ε*)-maxima) have been plotted. From the gradient of the line for the γ-process a value of $Q_{\gamma} = 60{,}7$ kJ/mol is obtained for the activation energy. In contrast, the activation energy of the glass transition is temperature dependent so that a single value cannot be given.

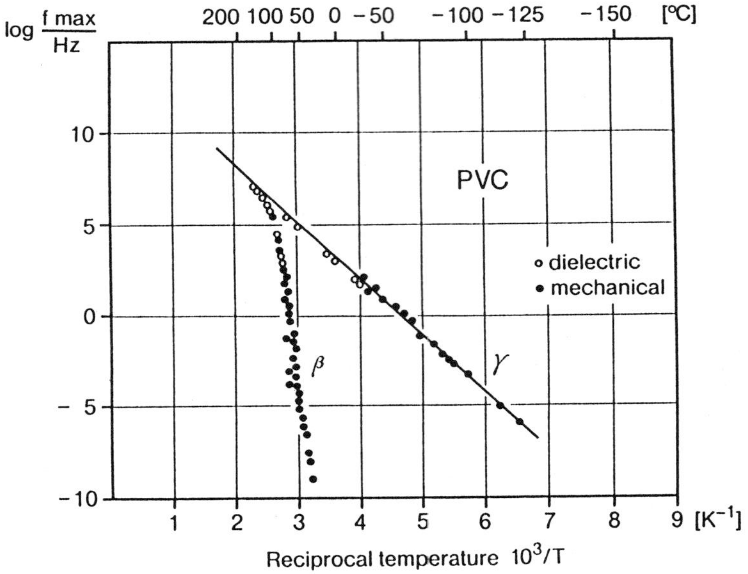

Fig. 24. Activation diagram for PVC
Mechanical data (from own measurements and [11]). Dielectric data (from [12])

4.5 Kink Model [13]

Polymers are composed of long chain molecules which oppose, with considerable strength, extension due to changes in the distance separating individual atoms and bending arising from changes in the bond angles along their backbones. If no other possibilities are available for movement then polymers would take on the properties normally associated with metals. However, there is the additional possibility of rotation around a C-C bond which involves only a relatively low potential threshold. Generally, a C-C bond has a threefold rotational potential (see Fig. 108 in Sect. 10.4). The quantitative course of this potential (height of the potential threshold, angular position of the minima) depends on the particular polymer whereas the threefold nature of the potential is, in principle, the same for most polymers since it arises from the sp^3 hybridization of the C-atoms. A model for the molecular motions responsible for relaxation processes, based both on experimental and theoretical results, has been developed by W. Pechhold. According to this model the jumping of particular rotational isomers, the socalled "kinks", is responsible for the low temperature relaxation process (γ-process). A kink exists in a planar chain when two gauche positions (120° positions in the rotational potential of polyethylene) are separated by a C-C bond having a trans-configuration. Kinks give rise to the smallest possible lateral displacement of adjacent straight parts of a chain and do

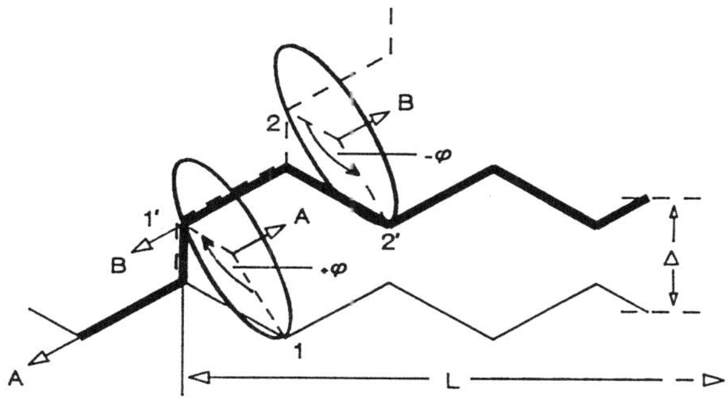

Fig. 25. Chain skeleton at a kink

not disturb the bundle as a basic structure. The formation of a kink is demonstrated in Fig. 25.

In an isotropic solid the kinks are equally distributed over energetically equivalent positions. An external stress causes a preferred occupation of one site and the kinks will jump to such a position. The change of site can be realised by a kink step or a kink jump (Fig. 26). Such motions resemble a crankshaft mechanism.

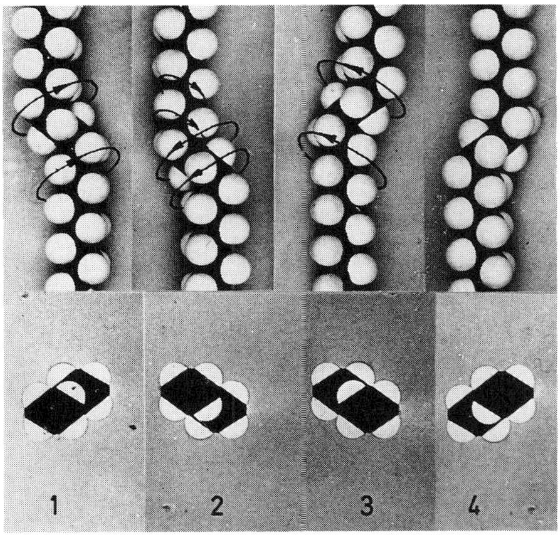

Fig. 26. Kink site changes in a molecular model of PE 1→2, 3→4 kink steps; 2→3 kink jump (From [13])

The rearrangement of a kink is connected with a paraelastic shear deformation, which has to be superimposed on the elastic deformation. In analogy to paramagnetism, the move to an energetically preferred position can be expressed by an excess probability which, in the equilibrium case, can be easily calculated from the Boltzmann factor.

This excess probability determines, together with the elastic compliance, the total deformation under external stress. From the paraelastic part of the total deformation the anisotropic relaxation strength can be obtained by differentiating with respect to stress. Subsequently, by spatial averaging, the macroscopic relaxation strength ΔJ of the γ-process can be obtained.

During a kink jump two gauche positions are exchanged so that twice the potential threshold associated with a rotation about a C-C bond is required. The trans-gauche potential threshold for polyethylene is 11.5 kJ/mol so that theoretically, the minimum activation energy for a kink jump is 23.0 kJ/mol. This compares well with the experimentally determined high temperature value for the γ relaxation process in polyethylene of 26.0 kJ/mol. Similary excellent agreement between theory and experiment have been obtained for a number of polymers.

5 Glass Transition [14–47]

5.1 Thermodynamic Description

The glass transition is a phenomenon which not only affects the modulus of polymeric materials; it also affects the specific volume, the enthalpy, the entropy, the specific heat, the refractive index, the dielectric constant etc., of such materials.

There are two basic theoretical descriptions of the glass transition:

a. Thermodynamic theory involving a second-order phase transition
b. Kinetic theory.

The slope of the curve of specific volume vs. temperature changes at the glass transition temperature. The exact position of this change of slope depends on the rate at which temperature is increased. Thus, the coefficient of thermal expansion, which is proportional to this slope, also changes rapidly at the glass transition (see Fig. 27). In analogy, at the temperature of the glass transition T_g, a rapid change in the specific heat c_p (the derivative of enthalpy with respect to temperature) and in the isothermal compressibility $\varkappa$ (the derivative of volume with respect to pressure) can be observed.

Formally, the experimental results suggest that vitrification is a second-order phase transition. For such transitions the Ehrenfest equations can be applied. Thus,

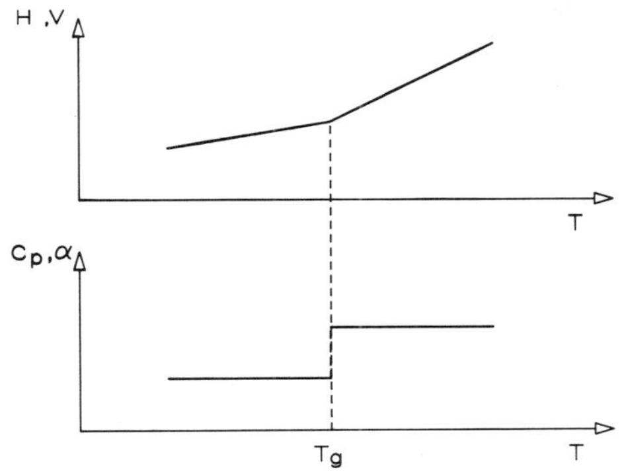

Fig. 27. Generalized temperature dependence of the enthalpy, the volume, the specific heat and the coefficient of thermal expansion at T_g

analogous to the Clausius-Clapeyron equation:

$$\left(\frac{dp}{dT}\right)_{tr} = \frac{\Delta H}{T\Delta V} \tag{99}$$

For a second-order phase transition this gives:

$$\left(\frac{dp}{dT}\right)_{tr} = \frac{\Delta\alpha}{\Delta\varkappa} \tag{100}$$

$$\left(\frac{dp}{dT}\right)_{tr} = \frac{\Delta c_p}{T\Delta\alpha} \tag{101}$$

Combining Eqs. 100 and 101 one obtains:

$$T_g = \left[\frac{\Delta\varkappa\,\Delta c_p}{(\Delta\alpha)^2}\right]_{T=T_g} \tag{102}$$

where

$\alpha = \dfrac{1}{V_0}\left(\dfrac{\partial V}{\partial T}\right)_p$ Thermal expansion coefficient

$\varkappa = -\dfrac{1}{V_0}\left(\dfrac{\partial V}{\partial p}\right)_T$ Compressibility

$c_p = T\left(\dfrac{\partial s}{\partial T}\right)_p$ Specific heat

Here Δ designates the difference between the values of the quantities above and below T_g.

A thermodynamic equilibrium phase transition would, however, require that the phases are indeed in equilibrium above and below the glass temperature; the existence of an equilibrium glass remains to be demonstrated for a polymeric material. Gibbs and DiMarzio have developed an equilibrium theory, according to which, an equilibrium transition some 50 K below the observable glass transition occurs but this too, remains to be experimentally verified.

5.2 Free Volume Theory [15–21]

The glass transition exhibits certain kinetic characteristics in that it is affected by the rate of cooling or heating as well as by the duration of loading and the frequency of an applied stress (see Figs. 28 and 29).

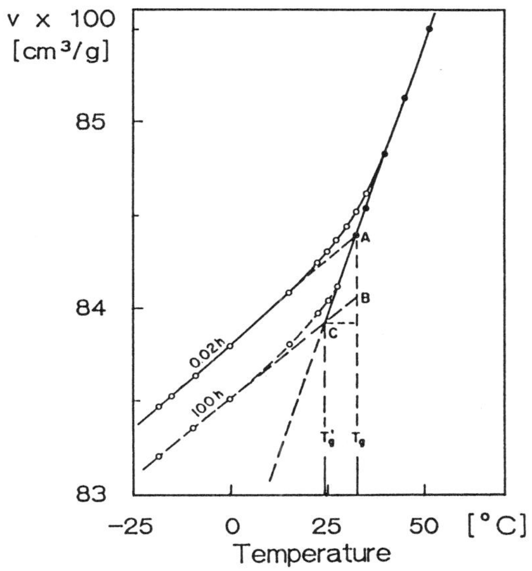

Fig. 28. Volume vs. temperature plots for polyvinyl-acetate at different rates of cooling (from [17]) *Closed circles:* equilibrium values

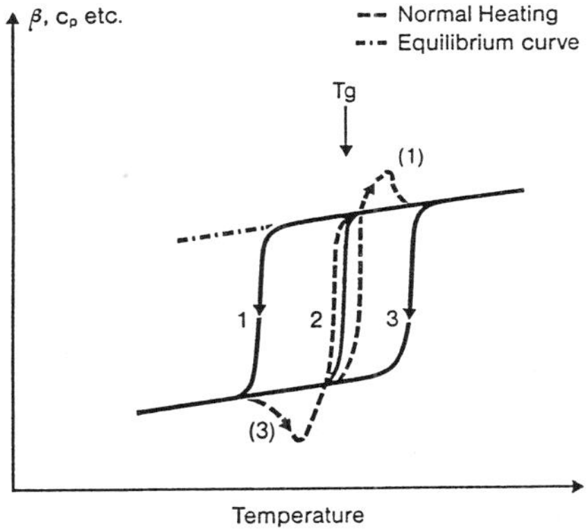

Fig. 29. Coefficient of thermal expansion and the specific heat as a function of temperature (from [15]). *1* low cooling rate < heating rate, *2* medium cooling rate = heating rate, *3* high cooling rate > heating rate

In this respect, the glassy state of polymers can be described thermodynamically as a non-equilibrium state. Such a description requires, in addition to the usual variables of state, an internal order parameter. This order parameter, according to Fox and Flory, is the socalled free volume, the definition of which is given in Fig. 30.

At the glass transition the molecular mobility is so drastically reduced that a non-equilibrium state would become frozen. This is evidenced by a constant free volume, which, according to Fig. 30, can be quantified by the difference between the total volume v and the molecular volume v_m, where the latter is the sum of a hypothetical volume in a melt free of voids at absolute zero and a volume expanded due to thermal vibrations v_s:

$$v_f = v_g + v_m \Delta\alpha (T - T_g) \tag{103}$$

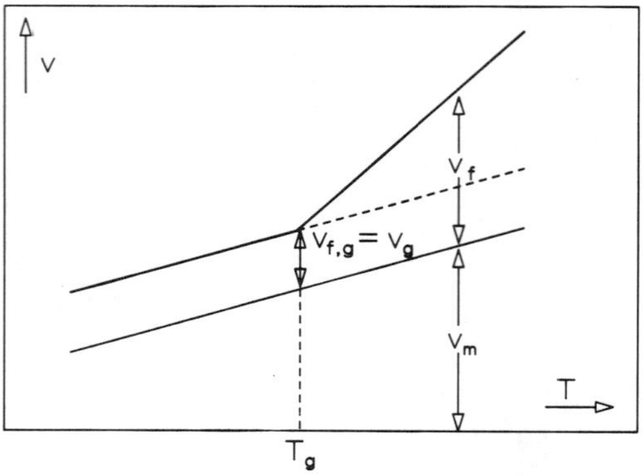

Fig. 30. Definition of free volume v_f and its dependence on temperature

At T_g the void volume becomes so small that cooperative jump processes by large chain segments, within the time-scale of measurement, can no longer occur; the material becomes "frozen". The probability that the voids are large enough ($v >> v^*$) for thermally activated jump processes to take place can be expressed, albeit simply, by the term $\exp -(v^*/v_f)$. As a consequence, the jump rate (Eq. 78) needs to be modified:

$$\Gamma = \text{const.} \exp(-Q^*/RT) \exp(-v^*/v_f) \tag{104}$$

From Eq. 104 the term for the energy of activation is given by:

$$Q(T) = Q^* + RTv^*/v_f \tag{105}$$

Here Q^* accounts for the temperature dependent amount of energy required to overcome the intramolecular potential barrier. Equation 105 provides an explanation for the temperature dependency of the Arrhenius curve of the glass transition process (see Fig. 24). Thus, as the temperature rises, the probability of enough void volume being available for a jump process increases. As a consequence, according to Eq. 105, the activation energy of the process decreases until, at a high enough temperature, the limiting lower value of Q^* (corresponding to the intramolecular potential threshold) is reached. Generally, this high temperature value agrees with the activation energy of the γ relaxation process so that it can be concluded that the mobility within both processes involves a similar mechanism.

5.3 Williams, Landel and Ferry Relationship [22, 23]

The origin of this approach lies in a model of vicosity based on the theory of free volume developed by Doolittle (see also Eq. 94):

$$\eta = B \exp(\beta v^*/v_f) \tag{106}$$

Here v* is the void volume required in order that a jump-activated chain segment can indeed make the jump and β is an empirical constant often approximated by 1. Putting η_1 and η_2 equal to the viscosities at temperatures T_1 and T_2 respectively:

$$\eta_1/\eta_2 = \exp\left[(v^*/v_{f1}) - (v^*/v_{f2})\right] \tag{107}$$

Furthermore, the free volume is defined by (see Eq. 103):

$$v_{f2} = v_{f1} + v_m \Delta\alpha (T_2 - T_1); \quad \text{Here} \quad v_m(T_1) \approx v_m(T_2) = v_m \tag{108}$$

Taking the logarithm of Eq. 107, inserting Eq. 108 and rearranging one obtains:

$$\ln \frac{\eta_1}{\eta_2} = \frac{(v^*/v_{f1})(T_2 - T_1)}{(v_{f1}/v_m \Delta\alpha) + (T_2 - T_1)} \tag{109}$$

For $T_1 = T_g$ Eq. 109 becomes:

$$\ln \frac{\eta_g}{\eta} = \frac{(v^*/v_g)(T - T_g)}{(v_g/v_m \Delta\alpha) + (T - T_g)} \tag{110}$$

On the basis of viscosity measurements at various temperatures the terms of the R.H.S. of Eq. 110 can be given the following numerical values:

$$v^*/v_g \approx 40 \quad \text{and} \tag{111}$$

$$v_g/v_m \Delta\alpha \approx 52 \tag{112}$$

The molar free volume necessary for a chain segment jump is some forty times the molar free volume at T_g. Using the approximation that the volume of activation v* is equal to the molecular volume of a chain segment v_m and applying the empirical relationships 111 and 112 for the difference between the expansion coefficients Δα above and below T_g and the fractional free volume f_g at T_g one obtains:

$$\Delta\alpha = \alpha_m - \alpha_g = 4{,}8 \cdot 10^{-4} K^{-1} \tag{113}$$

$$f_g = v_g/v_m \approx v_g/v^* = 0{,}025 \tag{114}$$

By taking logarithms to the base ten Eq. 110 can now be rewritten:

$$\log \frac{\eta(T)}{\eta(T_g)} = \log a_T = -\frac{17{,}44(T - T_g)}{51{,}6 + (T - T_g)} \tag{115}$$

Equation 115 is valid for all polymers assuming:

a) that the free volume at T_g is 2.5% of the total polymer volume and

b) that the change in the coefficient of thermal expansion $\Delta\alpha = 4.8 \cdot 10^{-4}\,K^{-1}$. Equation 115 is called the WLF-equation (After *W*illiams, *L*andel and *F*erry) and its general form is:

$$\log a_T = \frac{C_1(T - T_s)}{C_2 + (T - T_s)} \tag{116}$$

Usually, one takes as reference temperature T_s, a temperature some 50 K above T_g. Williams, Landel and Ferry have shown that a large number of polymers and polymer solutions as well as both organic and inorganic glasses can be described by Eq. 116 in the temperature range $T_s < T < T_s + 50\,K$ with $C_1 = -8.86$ and $C_2 = 101.6$.

5.4 Time-Temperature Superpositon Principle [24, 25]

The quantity $\eta(T)/\eta(T_g) = a_T$ is called the shift factor, the temperature dependence of which, according to Eq. 115, is plotted in Fig. 31.

Since most theories predict a direct proportionality between viscosity and relaxation time (see e.g. Rouse theory in 6.4 and Eq. 143), the shift factor a_T can be written:

$$a_T = \frac{\eta(T)}{\eta(T_g)} = \frac{\varrho(T)T\tau(T)}{\varrho(T_g)T_g\tau(T_g)} \approx \frac{\tau(T)}{\tau(T_g)} \tag{117}$$

In general, real material behavior cannot be described by a single relaxation time but rather by a continuous spectrum of relaxation times. If, however, the relative change in the jump frequency with temperature is the same for all relaxation times, then the

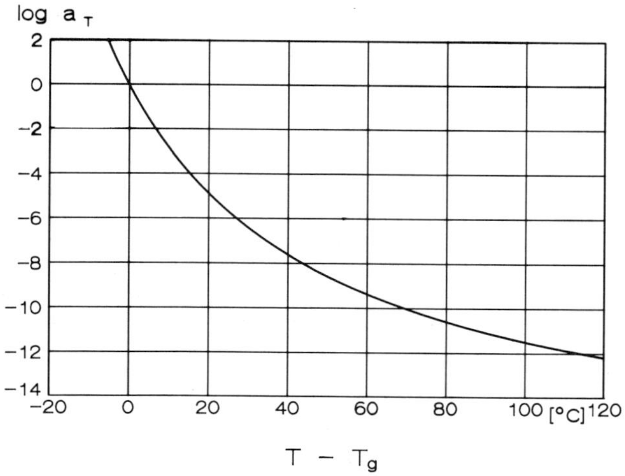

Fig. 31. WLF shift factor a_T as a function of temperature $(T - T_g)$ calculated from Eq. 115.

material is designated as "thermo-rheologically simple". In this case the plots of time and frequency dependent variables (e.g., G(t), G′(ω), G″(ω)), and the distribution functions (H(ln τ)) which can be calculated from these, are simply shifted parallel to the logarithmic time or frequency axis without changing their form. Thus, plots constructed from measurements made at various temperatures within a narrow time range can be reduced by parallel shift into a single, "master curve" (method of reduced variables). That is, for a time or frequency dependent modulus function:

$$E_T(t) = \frac{\varrho T}{\varrho_R T_R} E_{T_R}(a_T t) \tag{118}$$

or

$$G'_T(\omega) = \frac{\varrho T}{\varrho_R T_R} G'_{T_R}(a_T \omega) \tag{119}$$

Here T_R is the reference temperature at which the time or frequency dependent modulus function is to be plotted. The shift factor which results from the WLF-equation (Eq. 115) is only valid at temperatures above T_g but the time-temperature superpositon principle is generally valid so that the method of reduced variables can also be applied to lower temperature relaxation processes. The factor $\varrho T/\varrho_R T_R$ in Eqs. 118 and 119, which can be derived from a statistical theory of rubber elasticity, leads to a vertical shift of the modulus functions on a logarithmic scale. This factor is often set equal to 1 but can also be determined empirically. In Fig. 32 the principle of time and temperature superposition is demonstrated for polyisobutylene (PIB or Poly(methyl propene)).

Further examples of the application of this principle are given in Figs. 33 and 34. Figure 33 shows the shear modulus curves at the glass transition of PVC and Fig. 34

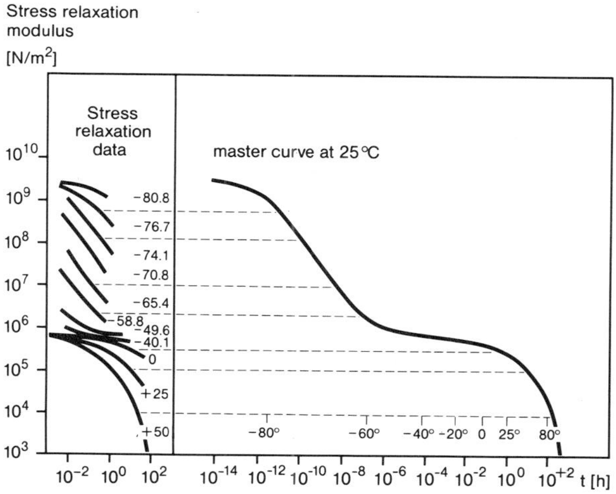

Fig. 32. Use of the time temperature superposition principle for PIB with $T_R = 25°C$ as reference temperature (from [25])

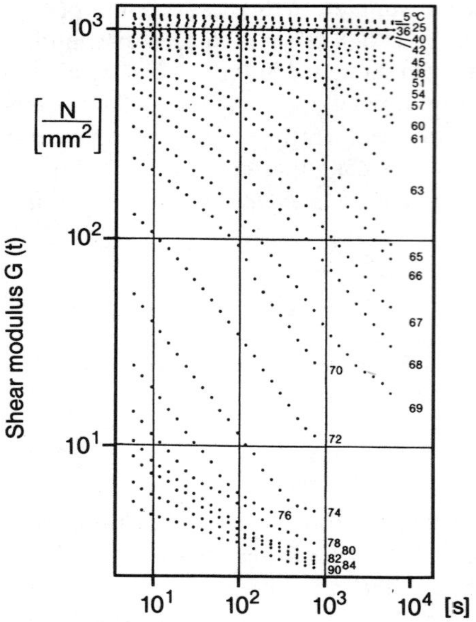

Fig. 33. Time dependent shear modulus curves for various temperatures in the region of the glass transition of PVC

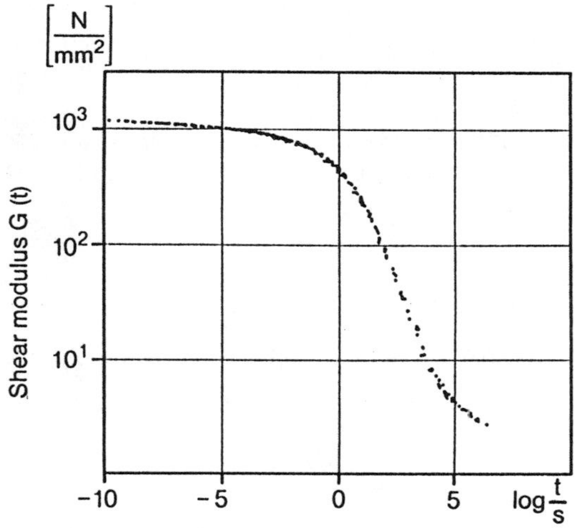

Fig. 34. Time dependency of the shear modulus of PVC; master curve for $T_g = 65\,°C$

is the master curve which can be derived from these measurements by shifting the modulus curves parallel to the logarithmic time axis using a reference temperature $T_R = 65\,°C$.

The shift factor calculated from time dependent relaxation measurements corresponds to that for variable frequency dynamic measurements: together they yield an Arrhenius diagram (see Fig. 24) from which, according to Eq. 98 and taking account of the free volume theory, the frequency shift factor is given by:

$$\log a_T = \frac{\log e}{R}\left[\frac{Q_g}{T_g} - \frac{Q(T)}{T}\right] \qquad (120)$$

Neglecting the temperature dependent part of the intramolecular potential Q* from Eq. 105, Eq. 120 becomes:

$$\log a_T = \log e \frac{v^*/v_g(T - T_g)}{v_g/v_m\Delta\alpha + (T - T_g)} \qquad (121)$$

With $v^* \approx v_m$ this can be simplified to:

$$\log a_T = \frac{1}{2{,}303\ f_g} \frac{T - T_g}{f_g/\Delta\alpha + (T - T_g)} \qquad (122)$$

Here $f_g = v_g/v_m$ corresponds to the fractional free volume at T_g. With $f_g = 2.5\%$ and $\Delta\alpha = 4{,}8 \cdot 10^{-4}\,K^{-1}$ Eq. 122 becomes the WLF equation (Eq. 115). Thus, it becomes clear that the time and frequency shift factors are self-consistent and can both be used to determine the activation energy of relaxation processes.

5.5 Increment Method for the Determination of the Glass Transition Temperature [26, 27]

Many authors have attempted to derive a relationship between the glass transition temperature and the polymer's chemical structure. The simplest method involves the calculation of T_g from the sum of increments ascribed to structural entities in the repeating units. With a proper choice of these subunits it is often possible to derive the glass transition temperature in terms of the individual contributions. In general:

$$T_g = \frac{\sum_i s_i T_{gi}}{\sum_i s_i} \qquad (123)$$

where T_{gi} is the contribution to the T_g of the structural entity i and s_i its weighting factor. The T_g-increments can be derived by regression analyses of T_g values from polymers of known structure. Hayes assumed that $\sum_i s_i T_{gi}$ corresponds to the molar cohesion energy. A more practical, modified increment method for calculating glass transition temperatures has been developed by van Krevelen and Hoftyzer. According to this method the glass transition temperature is given by:

$$T_g = \frac{Y_g}{M} = \frac{\sum\limits_i Y_{gi}}{M} \tag{124}$$

with Y_g as the socalled molar glass transition function. In Eq. 124 it is assumed that the Y_{gi} values of structural groups within any unit are independent of the nature of the adjacent moieties so that a simple additivity principle can be applied.

However, since this assumption does not always hold true, especially for polar moieties, correction terms can be introduced to allow for the interactions:

$$Y_g = \sum_i Y_{gi} + \sum_i Y_g(I_{xi}) \tag{125}$$

The interaction factor I_x is a measure of the concentration of polar groups in, for example, linear aliphatic polycondensation products such as polyesters, polycarbonates or polyamides.

As an example, one obtains the following group contributions to Y_g, using this method, for polyethylene terephthalate (from [27]):

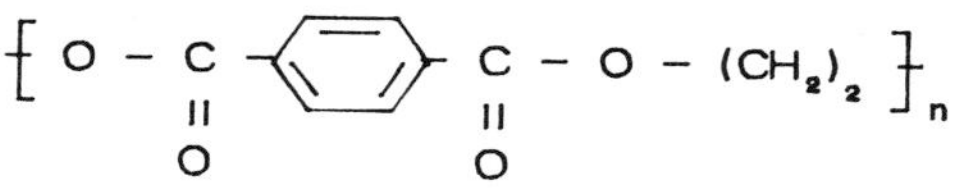

Group	Number	Y_{gi}
$-CH_2-$	2	5400
$-COO-$	2	16000
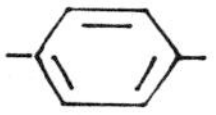	1	32000
$Y_g(I_x)$	2	12000
Y_g		65400

With M = 192 it follows (Eq. 124) that:

$$T_g = \frac{65400}{192} = 341\ \text{K}$$

Experimental values for such materials give values between 342 and 350 K. A large number of polymers give similarly good agreement so that this method is often used in industrial research to estimate the glass temperatures of target polymers. It should, however, be emphasized that this method is a deductive one which requires experimentally determined values from known structures.

5.6 Glass Transitions of Copolymers [28–31]

Because of the technological importance of copolymers a considerable effort has been expended with the aim of predicting the glass transition temperatures for such materials. Kelley and Bueche have developed a simple description of copolymer glass transition temperatures on the basis of the free volume theory. The additivity of the free volumes of two homopolymers leads to the following expression for the total fractional free volume of a copolymer in terms of the proportions of the two monomers:

$$f = 0{,}025 + \Delta\alpha_1\phi_1(T - T_{g1}) + \Delta\alpha_2\phi_2(T - T_{g2}) \qquad (126)$$

Here, ϕ_1 and ϕ_2 are the volume fractions of the homopolymers 1 and 2 respectively. Putting $T = T_g$ gives $f = 0.025$ so that:

$$T_g = \frac{\Delta\alpha_1\phi_1 T_{g1} + \Delta\alpha_2\phi_2 T_{g2}}{\Delta\alpha_1\phi_1 + \Delta\alpha_2\phi_2} \qquad (127)$$

and putting $\Delta\alpha_2/\Delta\alpha_1 = c$ one obtains:

$$T_g = \frac{\phi_1 T_{g1} + c\phi_2 T_{g2}}{\phi_1 + c\phi_2} \qquad (128)$$

A similar equation was developed by Gordon and Taylor phenomenologically and formulated, with $c \approx 1$, as:

$$T_g = \sum_i w_i T_{gi} \qquad (129)$$

Fox has also developed an approximate formula which he wrote in the form:

$$T_g = \frac{1}{\sum_i w_i/T_{gi}} \qquad (130)$$

In these equations w_i are the weight fractions of the individual components in copolymers or polymer mixtures. Figures 35 and 36 give an impression of how accurate Eq. 129 and 130 are when applied to statistical copolymers of acrylonitrile-butadiene (Fig. 35) and styrene-butadiene (Fig. 36).

However, considerable deviations from the theoretical values of free volume theory occur if one or more of the copolymer's components is crystalline. Thus, for example, the hydrogenation of poly(acrylonitrile-co-butadiene) leads to crystallizable ethylene sequences and the glass transition temperatures of the products are considerably higher than would be expected for the amorphous copolymers (see Fig. 37).

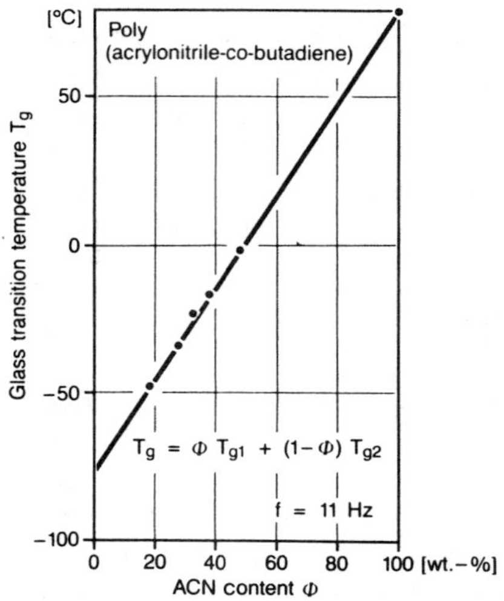

Fig. 35. T_g of statistical acrylonitrile-butadiene co-polymers as a function of ACN-content

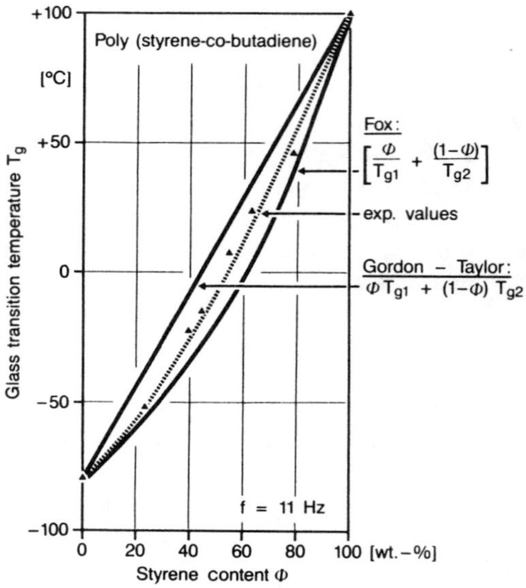

Fig. 36. T_g of statistical styrene-butadiene copolymers as a function of styrene content

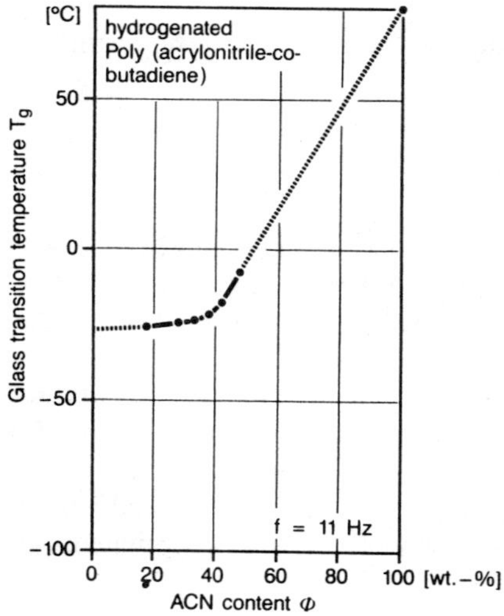

Fig. 37. T_g of hydrogenated acrylonitrile-butadiene copolymers as a function of ACN content

5.7 Dependence of T_g on Molar Mass [18, 32, 33]

As long ago as 1950 Fox and Flory proposed the relationship

$$T_g = T_{g\infty} - \frac{K}{M} \tag{131}$$

between glass transition temperature and molar mass. This equation is also based on free volume theory. Thus, every free chain end has a larger free volume than an identical structural unit incorporated within a long chain and a polymer with more free ends must be cooled to lower temperatures in order to have the same free volume as a polymer with fewer free ends. This implies that with decreasing molar mass the glass transition is shifted to lower temperatures. With:

v_e = excess free volume per chain end,
$2v_eN_L$ = excess free volume per Mol and
$2\varrho v_eN_L/M$ = excess free volume per cm^3

and assuming that the free volume is constant and independent of molar mass at the glass transition temperature, one obtains:

$$2\varrho N_L v_e/M = \alpha_f(T_{g\infty} - T_g) \tag{132}$$

In Eq. 132 α_f is the expansion coefficient of the free volume and can be approximated by the difference $\Delta\alpha$ between the expansion coefficients of the glass and the melt. It follows from Eq. 132 that:

$$\begin{aligned} T_g &= T_{g\infty} - 2\varrho N_L v_e/\Delta\alpha\, M \\ &= T_{g\infty} - K/M \end{aligned} \tag{133}$$

In these equations T_g is the glass transition temperature for a polymer with an infinite molar mass. The limiting molar mass, above which the T_g is essentially constant, is usually between 10^4 and $5 \cdot 10^4$ Daltons. Furthermore, if the value of K in Eq. 133 is determined experimentally then v_e can also be obtained. For example, for polystyrene Beevers and White have found $v_e = 40 Å^3$. Kanig and Ueberreiter have discussed an equation which is equivalent to Eq. 133, namely:

$$1/T_g = 1/T_{g\infty} + K'/M \tag{134}$$

However, a considerably improved agreement between experiment and theory is achieved by using the modified Fox-Flory equation developed by R. Becker:

$$T_g = T_{g\infty} - C/M^a \tag{135}$$

in which C and a are empirical constants.

5.8 Empirical Correlations Between Molecular Parameters and Glass Transition Temperatures [15, 34–47]

One molecular parameter which exerts an important influence on the glass transition temperature is the stiffness of the polymer backbone and this can be correlated with the cross-section of the backbone. The temperature dependent shear modulus measurements made by Schmieder and Wolf on polymers with various cross-sections are shown in Fig. 38.

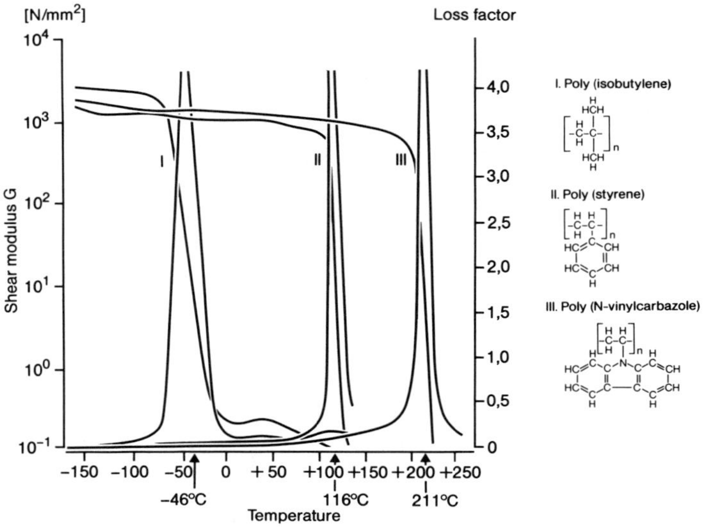

Fig. 38. Temperature dependency of the shear modulus and of the logarithmic decrement for several polymers as measured by Schmieder and Wolf (from [34])

From the glass transition temperatures of the aromatic vinyl polymers listed in Table 3 and from Fig. 38 it can be concluded that larger substituents on the polymer backbone stiffen the chain and shift the glass transition to higher temperatures. This can be rationalized in terms of the free volume model: The larger the substituent the lower is the probability that the void volume exists, which is required to enable a thermally activated jump of the corresponding segment.

Table 3 is a collation of some glass transition temperatures for a selection of aromatic vinyl polymers.

There are, however, examples which appear to contradict this conclusion (see Table 4).

The disubstituted polymers in Table 4 have lower glass temperatures than their monosubstituted counterparts. Nevertheless, it must be remembered that disubstitution leads to an approximately cylindrical polymer chain and that such molecules have a smaller packing density.

Table 3. Glass transition temperatures of some aromatic vinyl polymers (from [15])

Polymer	Structural formula	T_g
Polystyrene	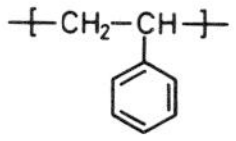	100
Poly-o-methyl styrene	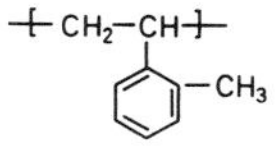	115
Poly-1-vinyl naphthalene	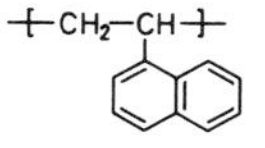	135
Poly-vinyl biphenyl	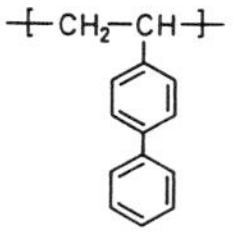	145
Poly-α-methyl styrene	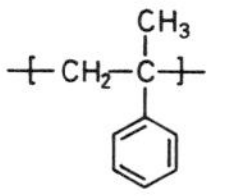	175
Poly-acenaphthalene	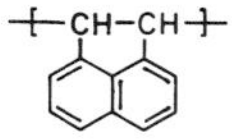	264

Table 4. Glass transition temperatures of some polymers with main chain mono- and di-substution

Polymer	Structural formula	T_g
Polyvinylchloride	$-(CH_2\text{-}CHCl)-$	+87
Polyvinylidene chloride	$-(CH_2\text{-}CCl_2)-$	−17
Polypropylene	$-(CH_2\text{-}CHCH_3)-$	−10
Polyisobutylene	$-(CH_2\text{-}C(CH_3)_2)-$	−65

Another important influence on the glass transition is exercised by side chains and their mobility. Whereas side groups which stiffen the main chain increase the transition temperature, flexible side chains lead to a decrease in T_g (see Table 5).

Table 5. Glass transition temperatures of some polymers having butyl side chains (from [15])

Polymer \ Bu	$-C(CH_3)_2-CH_3$	$-CH(CH_3)CH_2CH_3$	$-CH_2CH_2CH_2CH_3$
$+CH_2-CH(Bu)+$	59	36	-36
$+CH_2-CH(COOBu)+$	43	-22	-56
$+CH_2-C(CH_3)(COOBu)+$	–	53	21
$+CH_2-CH(C_6H_4Bu)+$	118	–	6

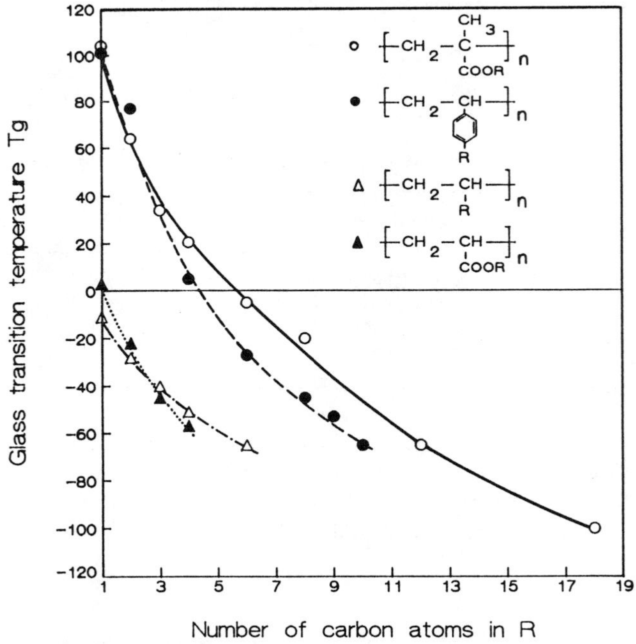

Fig. 39. Glass transition temperature as a function of side chain length (from [15])

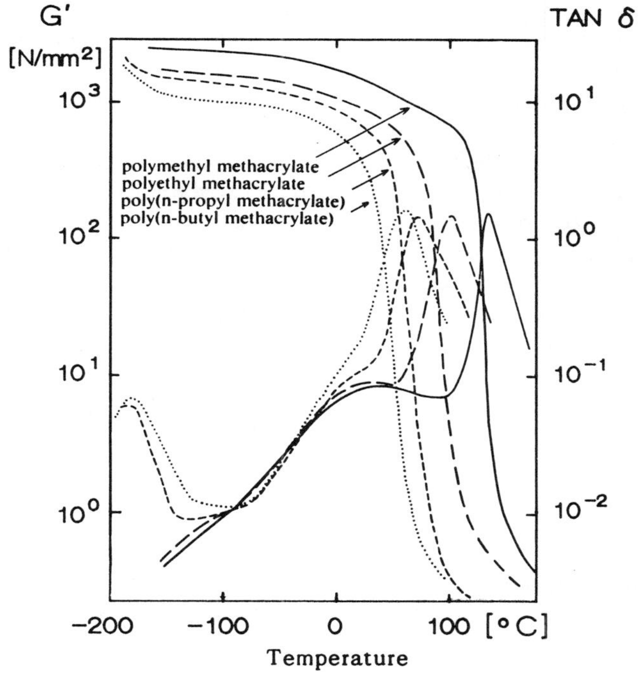

Fig. 40. Temperature dependence of the shear modulus and the damping for some poly (*n*-alkyl methacrylates) (from [35])

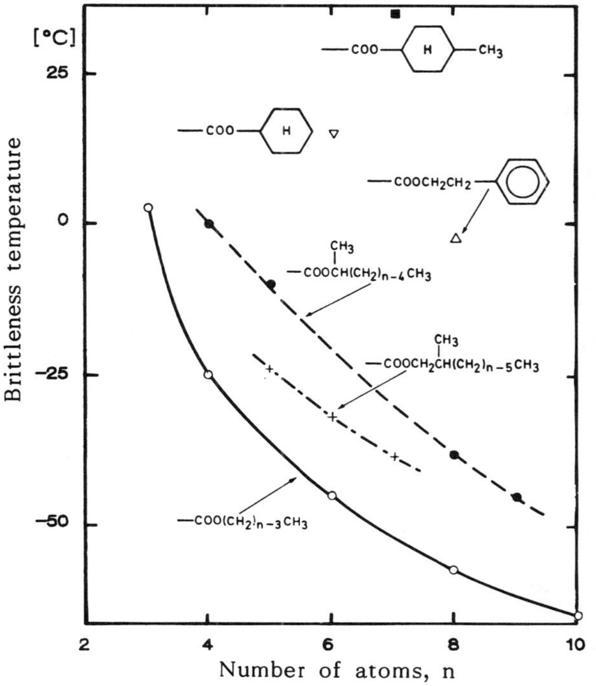

Fig. 41. Dependence of the brittleness temperature on the length of the polymer side-chains (from [35])

For the polymers used as examples in Fig. 39, 40 and 41 the T_g is shifted to lower temperatures as the length of the side chain increases: The flexible side chains act as an intermolecular "diluent" and thus increase the free volume. In contrast, if the side-chain is long enough to be able to crystallize, this effect is ameliorated or even reversed and the glass temperature increases. Figure 41 shows that it is the flexibility of the side chain rather than simply the length that determines the glass transition temperature. Thus, a stiffening of the side-chain due to a methyl branch 3 carbon atoms removed from the backbone leads to an inrease in the glass transition temperature of ca. 20 K; the same structure 4 carbon atoms removed from the backbone inreases the glass transition temperature by only ca. 10 K. A chain stiffening cyclohexyl group increases the glass transition temperature by ca. 60 K compared to that for an *n*-butyl group.

5.9 Plasticizer

Plasticizers lead to an effect on the glass transition temperature similar to that produced by flexible side-chains. The plasticizer molecules slide between the polymer chains leading to an increase in their free volume and a decrease in the

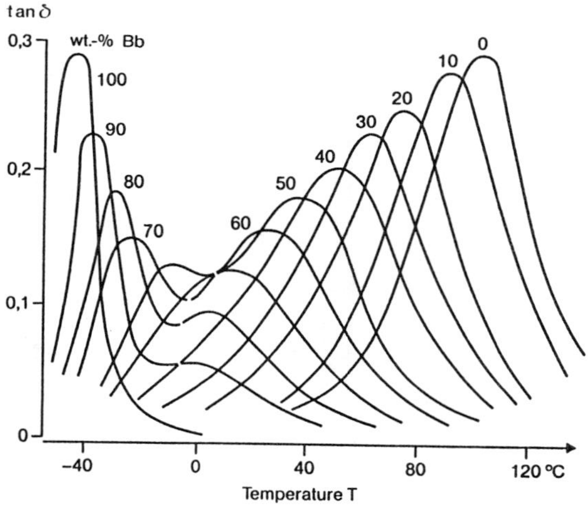

Fig. 42. Temperature dependence of the loss factor for polyvinyl acetate plasticised with benzyl benzoate (Bb) (from [36])

glass transition temperature of the polymer (see Fig. 42). Of course, in order to achieve this effect the glass temperature of the plasticizer must lie below that for the polymer and the plasticizer and polymer must be compatible (at least in the concentrations present).

5.10 Crosslinking

As can be seen from Fig. 43 the glass temperature increases rapidly when a natural rubber is vulcanized. In contrast to this picture, crosslinking with peroxide or radiation to give a similar crosslink density (equal modulus in terms of Fig. 44) leads to only a slight increase in the glass transition temperature. The large shift in the glass temperature for a NR-vulcanisate with increasing addition of sulfur can be attributed to an intramolecular cyclization involving S-atoms which reduces the mobility of the polymer backbone. An analogous shifting of the temperature or frequency at which the glass transition occurs can be observed for peroxide or radiation cured polymers but only above a certain level of crosslinking.

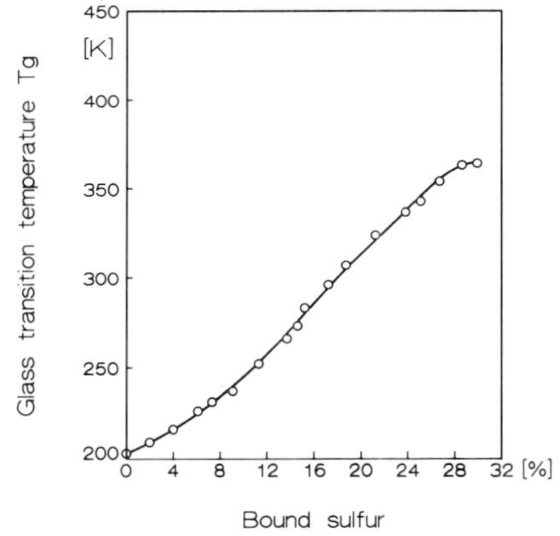

Fig. 43. Glass transiton temperature of NR as a function of crosslink density (from [15]). (Data from Martin and Mandelkern [43])

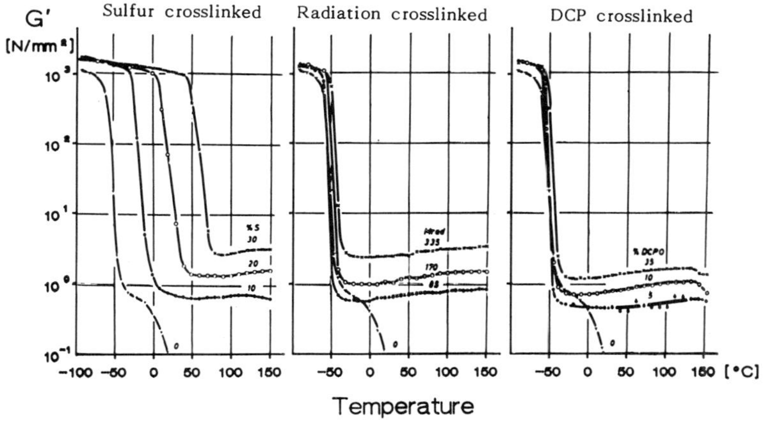

Fig. 44. Temperature dependence of the shear modulus of NR crosslinked by various methods to a range of crosslink densities (from [44])

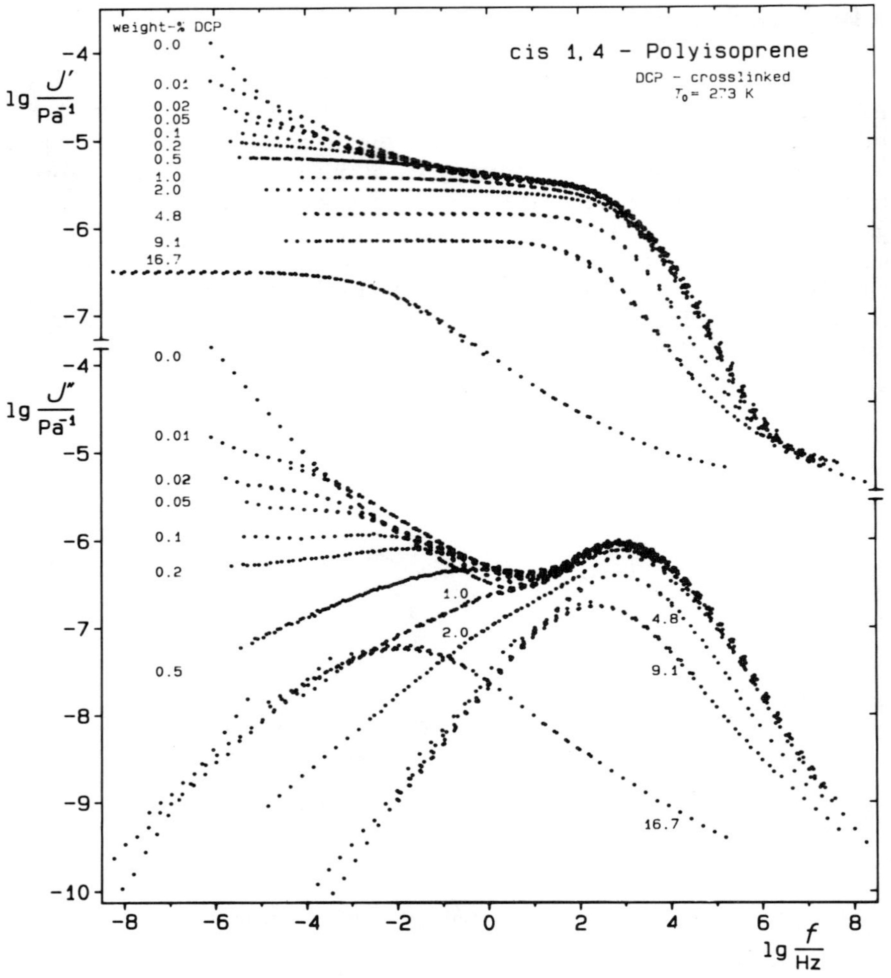

Fig. 45. Frequency dependence of the dynamic shear compliance of polyisoprene at different peroxide levels (from [45])

The measurements of dynamic shear compliance made by Pechhold and his coworkers on peroxide crosslinked polyisoprene with up to 9 wt.-% peroxide showed a lowering of the rubber plateau proportional to the amount of peroxide (DCP) whereas the frequency of the glass transition remained constant (see Fig. 45). An increase in the amount of peroxide from 9.1 to 16.7 wt.-% led to a jump in the glass transition to a frequency some 5 decades lower. At this concentration of crosslinks the deformation process responsible for the glass transition is obviously strongly hindered.

5.11 Fillers

The temperature and frequency of the glass transition of a polymer is generally only slightly affected by the addition of fillers in technologically relevant amounts. In this respect the nature of the bonding of the filler to the polymer matrix is unimportant. The active carbon blacks, essential for the reinforcement of many elastomers, give rise to "bound rubber"; a layer of rubber some 3 nm thick, which is bound to the filler via physical and chemical bonds and is thus insoluble. Although, according to Kraus and Gruver, this immobilized layer should have a glass temperature some 10 K higher than the matrix a dependence of the glass transition temperature on the filler has not been observed in reinforced systems. As can be seen in Fig. 46, flow in filled systems is not suppressed by the crosslinks between the filler particles and the rubber matrix. Thus the sliding processes are enabled, which regulate the flow of the polymer chains within the matrix and probably also to a limited extent that along the filler-polymer interfaces.

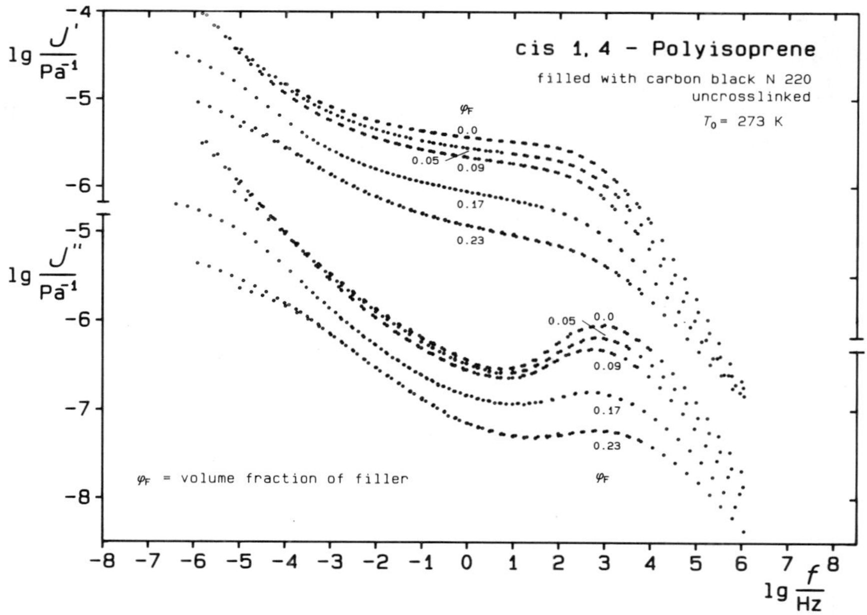

Fig. 46. Frequency dependence of the dynamic shear compliance of polyisoprene containing varying amounts of filler (from [45])

6 Flow and Rubber Elasticity in Polymer Melts [4,45, 46, 48–57]

6.1 Flow as a Relaxation Process

In addition to those relaxation processes which have been discussed up to now (γ-process, glass transition) there is an additional process, which occurs in all uncrosslinked polymers: the flow relaxation process. By this process the polymer melt exhibits a recoverable strain if the stress is released. In dynamic mechanical investigations, this corresponds to a paraelastic relaxation process taking place over the same frequency range as that where $J''=1/\omega\eta$. This process occurs on the temperature scale above the glass transition, whereas on a frequency scale it occurs below this transition. Figure 47 shows, as an example, the frequency dependent master curves of the dynamic shear compliance and the shear modulus for polyisobutylene at a reference temperature, $T_0=273\,K$, obtained using the

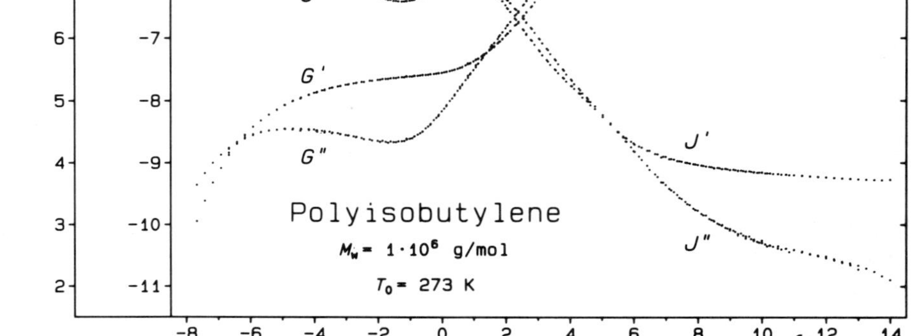

Fig. 47. Frequency dependence of the dynamic shear compliance and the shear modulus for polyisobutylene (from [45])

frequency-temperature superposition principle. In this graph one can see (from *left* to *right*): flow, the rubber plateau, the glass transition and the modulus and compliance plateaus of the glassy, frozen solid. From a comparison of the two curves it is obvious that the maximum loss modulus at the glass transition occurs at a frequency some six orders of magnitude greater than that of the loss compliance. The modulus plateau (10^0 to 10^{-4} Hz) at the low frequency end of the glass transition is ascribed to physical crosslinks in the melt (e.g., entanglements of longer chain-molecules), which give rise to rubber elastic behavior analogous to that produced by covalent crosslinks. A further frequency reduction leads to the flow process, whereby long-range segment movements are activated. In the limiting case of very long time ($\omega \to 0$) flow dispersion becomes irreversible flow; the latter can be described by the steady flow viscosity ($\approx$Newtonian viscosity)

$$\eta_0 = \lim_{\omega \to 0} (G''/\omega) = \lim_{\omega \to 0} (1/\omega J'') \tag{136}$$

6.2 Structural Models for Polymer Melts

In order to understand the flow behavior of polymer melts in terms of the individual molecules a knowledge of the supramolecular structure of such melts is required. At the time of writing there are essentially two basic, mutually exclusive, models for the molecular order in a polymer melt. The model describing the melt in terms of statistically coiled, interpenetrating and entangled polymer molecules has achieved considerable acceptance (Fig. 48, *left*). In this model local order occurs only in the smallest regions (of the order of 5 nm) and such regions are continually affected by thermal fluctuations. An alternative, the meander model was developed by W. Pechhold for polymer melts and amorphous solids (Fig. 48, *right*).

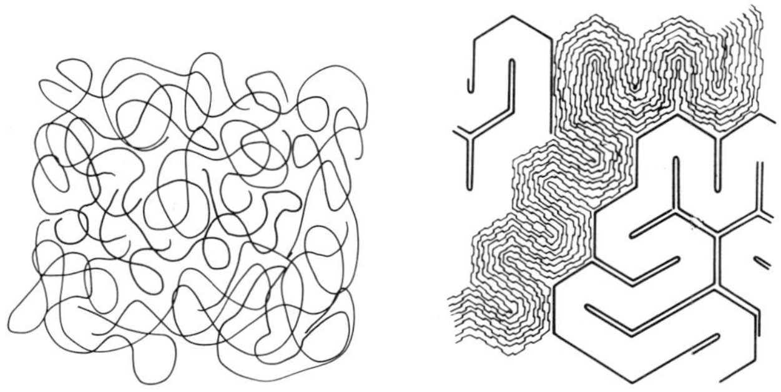

Fig. 48. Coil (*left*) and meander (*right*) models of a polymer melt

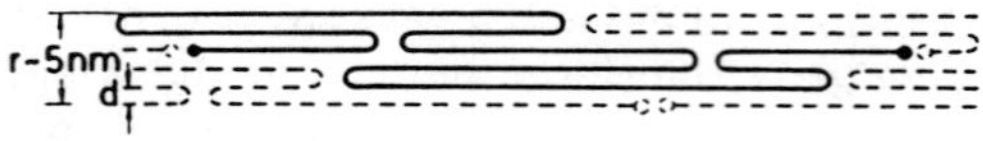

Fig. 49. Melt bundle of molecules (1 to 10 chains depending on M) (from [48])

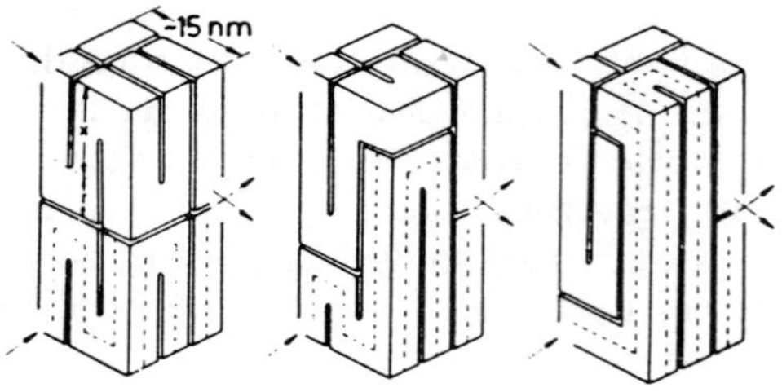

Fig. 50. Superfolding of melt bundles into 9-fold meander cubes and x/r fluctuations (from [48])

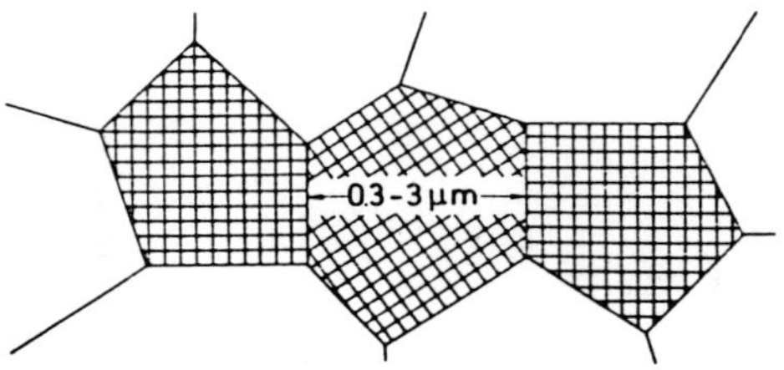

Fig. 51. Aggregation of meander-blocks into isotropic coarse grains (from [48])

Pechhold's argument against the coil model is based on the not inconsiderable difficulty, particularly for stiff chains, of obtaining optimal spatial packing without assuming large, elastic chain deformations. The basis of the meander model is a bundle of molecules in their energetically most favorable conformation and in which all thermodynamically reasonable disorders are allowed. These disorders are produced by a well-defined class of rotational isomers (kinks), which leave the molecules essentially stretched so that a bundle can still be formed. Since molecule bundles form anisotropic structures they cannot fill a volume isotropically; "knick areas" are introduced which alter the orientation of the chain bundles and increase the orientational entropy of the segments. Tight superfolding of the molecule bundles then allows cubic subunits to be constructed and these form superstructures which isotropically fill the available volume.

The basic elements of such superstructures are shown in Figs. 49–51. A melt bundle is composed of 3 to 10 (the number depends on the molar mass) statistically tightly folded and reptating molecules resulting in a one dimensional (quasiparallel) coil appearance of its overall geometry (Fig. 49). Superfolding of the topological bundle, in which the statistically folded chains are arranged lengthwise, leads to space-filling, cubic meander blocks, linked together via space diagonals as axes of meander cube rotation. The melt structure is formed by an aggregation of the meander blocks into coarse grains, whereby isotropy results from a statistical sequencing of the meander cubes, such that these occupy all three accessible sites. In liquid crystal polymers, identical cube positions are preferred and a high degree of orientation results.

Furthermore, the individual grains reflect the anisotropy of the meander blocks and structures, which can be observed under polarized light, may be identified as such.

The meander model is based on the socalled "Cluster Entropy Hypothesis" (CEH). This states that all types of cluster building within subspaces (e.g. conformation, orientation, deformation) involving m equivalent structural elements (e.g., segments, dislocations and layers), each having f accessible states (vibrational, conformational etc.) do not reduce the entropy provided $m < f$. With this assumption the free energy is minimized by contributions resulting from the folding of individual chains and the superfolding of molecular bundles. In this model reptational movements of the individual chains in a bundle, x/r and r/d fluctuations of the bundle (see Fig. 49 and 50), rotation of meanderblocks and shear deformations are all allowed. Using the meander model it has proved possible, for the first time, to quantitatively describe the total relaxation behavior of polymeric solids and melts, including the rubber elasticity of crosslinked systems (see also Sect. 6.6).

6.3 Bueche-Rouse Model [4, 49–51]

In 1953 P.E. Rouse developed a model and a theoretical description for flow in dilute solution. According to this model, the polymer molecule exists as a statistical coil and is subdivided into N submolecules so that the end-to-end distance for each submolecule can be described by Gaussian statistics. Furthermore, according to this model, the equilibrium configuration for each molecule can be described in terms of N vectors or as a point in a 3-dimensional phase space and the mass of each submolecule is thought of as being concentrated in a solid bead. The submolecules behave as Gaussian chains so that their entropy-elastic recovery can be described by a spring with a spring constant $3kT/a^2$ (a = average end-to-end distance of a submolecule, see Fig. 52).

If one considers only movements in the x-coordinate, the equation of movement for the ith bead is given by:

$$m\ddot{x}_i + k_0\dot{x}_i + 3kT/a^2(-x_{i-1} + 2x_i - x_{i+1}) = F_{xi} \quad (137)$$

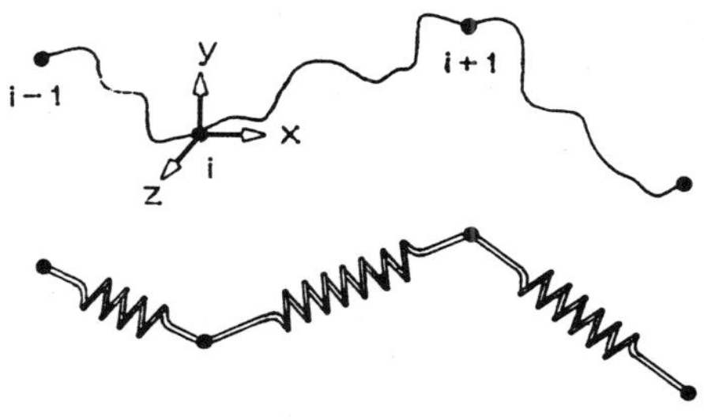

Fig. 52. Bead-spring model of a polymer chain as envisaged by Rouse

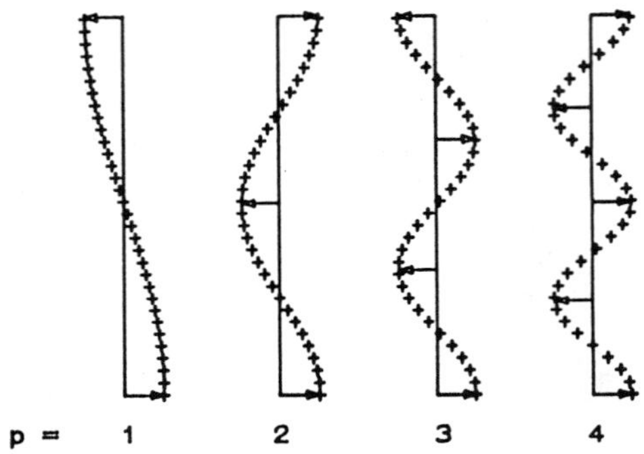

Fig. 53. Vibrational modes of an elastic rod

The component of inertial force is neglected since it is small compared to that due to friction. The Rouseian bead-spring model has been developed and applied to concentrated solutions and polymer melts by F. Bueche. In this development the solvent molecules which surround each polymer molecule in a dilute solution are substituted by equivalent polymer molecules. The coil form of the individual polymer molecules is assumed to be determined by its environment, as in dilute solution. The total loss of energy of individual polymer molecules, due to friction, as they move through the viscous polymer matrix is simply accounted for by the friction coefficient k_0, an attribute of each individual submolecule (Eq. 137). The movements of such a bead-spring molecule can be envisaged by considering them as analogous to those of a longitudinally vibrating elastic rod. Thus, a normal coordinate transformation leads to a series of N normal modes of motion with characteristic proper frequencies and relaxation times which correspond to those of an elastic rod (see Fig.53).

6.4 Rouse Theory of Flow in Low Molar Mass Polymer Melts [4]

Using the differential Eq. 137 one can obtain the resonance frequencies for the individual vibrational modes and also the visco-elastic material constants. In analogy to Maxwell's model one obtains for the frequency dependent, complex shear modulus (Ferry, [4]):

$$G' = (\varrho RT/M) \sum_{p=1}^{N} \omega^2\tau_p^2/(1 + \omega^2\tau_p^2) \tag{138a}$$

$$G'' = (\varrho RT/M) \sum_{p=1}^{N} \omega\tau_p/(1 + \omega^2\tau_p^2) \tag{138b}$$

$$G(t) = (\varrho RT/M) \sum_{p=1}^{N} e^{-t/\tau_p} \quad \text{with} \tag{139}$$

$$\tau_p = \frac{a^2N^2k_0}{6\pi^2p^2kT} \tag{140}$$

where ϱ is the density and M the molar mass. Thus, the flow dispersion is characterized by a discrete relaxation-time spectrum (Eq. 140), whereby each relaxation-time makes a constant contribution $\varrho RT/M$ to the visco-elastic modulus functions. For the viscosity:

$$\eta'(\omega) = \varrho RT/M \sum_{p=1}^{N} \tau_p/(1 + \omega^2\tau_p^2) \tag{141}$$

is valid, so that using Eq. 140 with $\sum_p (1/p^2) = \pi^2/6$

$$\eta_0 = \lim_{\omega \to 0} \eta'(\omega) = \frac{a^2N^2k_0n}{36} \tag{142}$$

Here $n = \varrho N_L/M$ gives the number of molecules/cm^3 with N_L = Avogadro's number.

In this way the relaxation times given by Eq. 140 can be described more usefully by:

$$\tau_p = \frac{6\eta_0 M}{\pi^2p^2\varrho RT} = \frac{\tau_{max}}{p^2} \tag{143}$$

In the terminal zone of the frequency scale the flow behavior of the polymer melt is dominated by the vibrational mode with the longest relaxation time: $\tau_1 = \tau_{max}$. This mode of motion corresponds to coordinated movement of the molecule as a whole. The terminal relaxation time is a measure of the time required to attain steady-flow viscosity under constant stress. In this frequency region the following approximations for the visco-elastic modulus functions can be used:

$$G'_{\omega \ll 1/\tau_1} = (\varrho RT/M)\omega^2\tau_1^2 \sum_{p=1}^{N} (\tau_p/\tau_1)^2 \tag{144a}$$

$$G'_{\omega \ll 1/\tau_1} = (\varrho RT/M)\omega\tau_1 \sum_{p=1}^{N} (\tau_p/\tau_1) = \omega\eta_0 \tag{144b}$$

Thus, in the low frequency region, Rouse's theory predicts a slope of 1 or 2 for a double logarithmic plot of the loss modulus or the storage modulus respectively. In addition, Rouse's theory suggests that the first terms in the summations of Eqs. 144a and 144b dominate the stationary values of the viscosity and the compliance. Thus, τ_1 contributes some 61% of the total viscosity (Ferry [4]):

$$\eta_0 = (\varrho RT/M)\tau_1 S_1 \quad \text{with} \quad S_1 = \sum_{p=1}^{N} (\tau_p/\tau_1) \tag{145}$$

and some 92% of the total compliance:

$$J_e^0 = G'/\omega^2\eta_0 = (M/\varrho RT)S_2/S_1^2 \quad \text{with} \quad S_2 = \sum_{p=1}^{N} (\tau_p/\tau_1)^2 \tag{146}$$

After the low frequency region of the Rouse spectrum (to higher frequencies, $\omega\tau_1 \approx 1$) there is a transition zone where both modulus functions assume a slope of 1/2:

$$G' = G'' = (\sqrt{3}/2)(\varrho RT\eta/M)^{1/2}\omega^{1/2} \tag{147}$$

The approximations are valid for polymers having uniform molar mass. From Eq. 142 it can be seen that the, socalled, monomer coefficient of friction ζ_0 (the frictional force per monomer unit per unit velocity, see Eq. 137) is proportional to η_0/M. Thus, measurements of viscosity can yield information about the coefficient of friction. The ratio η_0/M increases initially with increasing M and then assymptotically approaches a limiting value. These correlations are, however, only valid for molar masses where no molecular entanglements exist. The molar mass dependence of ζ_0 at low molar masses is ascribed to the additional free volume associated with the chain ends; the number of chain ends per unit volume is proportional to 1/M. Thus, the influence of the chain ends decreases with increasing molar mass and

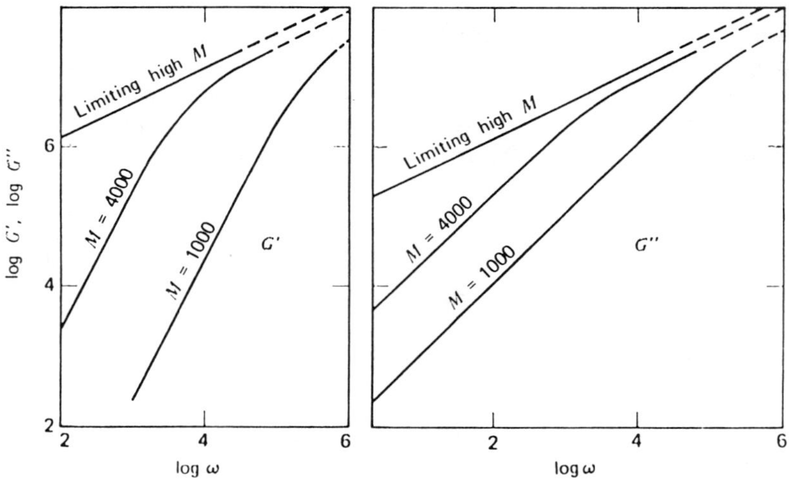

Fig. 54. Schematic representation of the complex shear moduli as functions of frequency and molar mass in the region of flow dispersion as predicted by Eqs. 143, 144a, 144b, 147, 148 and 149 (from [4])

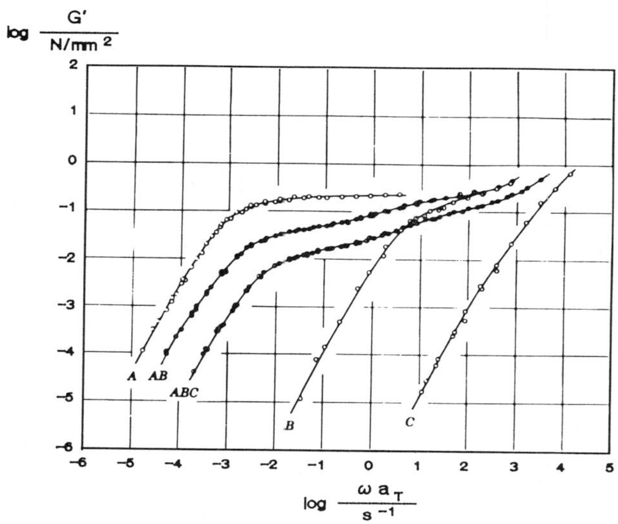

Fig. 55. Frequency dependence of the storage modulus reduced to 160°C for narrow molar mass distribution polystyrenes with various molar masses ($M_C < M_B < M_A$) and mixtures of these (from [4])

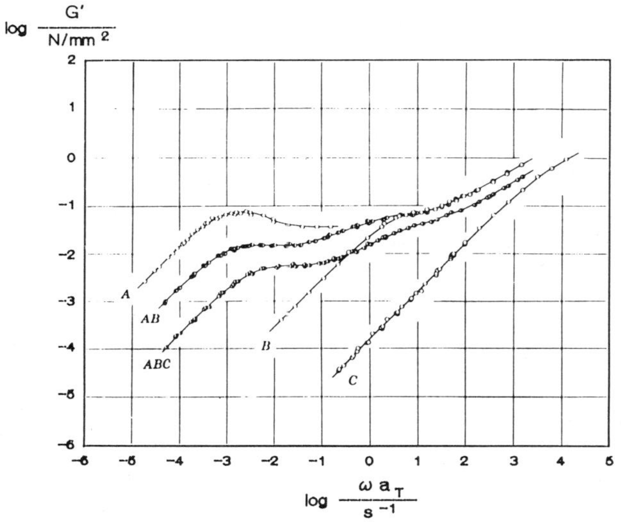

Fig. 56. Frequency dependence of the loss modulus reduced to 160°C for polystyrenes of various molar masses ($M_C < M_B < M_A$) and mixtures of these as in Fig. 55 (from [4])

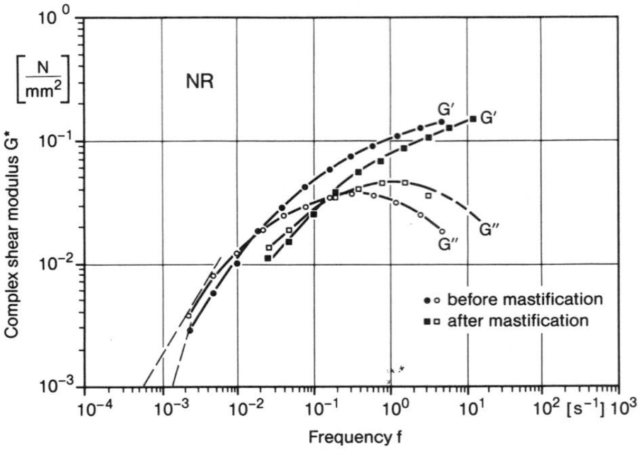

Fig. 57. Frequency dependence of the storage and loss moduli for NR before and after mastification on a two-roll mill

eventually becomes negligable. The theory of free volume according to Eq. 106 can be written:

$$\log \zeta_0 = \log \zeta_{00} + (\beta/2{,}303)(1/f_M - 1/f_0) \tag{148}$$

$$f_M = f_0 + A/M \tag{149}$$

where: f_M is the fractional free volume at molar mass M; f_0 and ζ_{00} are the corresponding limiting values at large molar mass; A is a polymer specific constant and β is a numerical constant close to unity. Figure 54 shows schematically the curves of the modulus functions G′ and G″ as a function of frequency in the region of flow dispersion.

The Rouse theory has shortcomings both for lower and higher frequencies. Additionally, Rouse's theory is only valid up to molar masses of ca. $M < 2 \cdot 10^4$. The theory does not take account of rubber elastic behavior between the flow region and the glass transition of large molar mass polymer melts. Furthermore, the number of vibrational modes which can be activated is limited by the number of submolecules. Since the submolecules should always form partial coils, their minimum size is limited. Thus, no frequency modes with frequencies greater than $\omega_N = 1/\tau_N$ can be activated since these require shorter submolecules.

From the preceding figures (Figs. 55, 56 and 57) it can be seen that the Rouse theory is corroborated by experiment, at least in the terminal zone of the frequency scale.

6.5 Extension of the Rouse Theory to Large Molar Mass and Crosslinked Polymer Melts [4]

The long range modes of motion to the low frequency side of the Rouse spectrum are hindered in large molar mass melts by entanglements. A formal trick can, however, be applied to allow a description of the behavior over longer time scales: The vibrational modes which are hindered for entangled molecules can be treated as more strongly hindered and, thus, to be moving in a more viscous medium than the other molecules. To this end the monomer coefficient of friction ζ_0 is substituted by $Q_e\zeta_0$, where Q_e is a numerical factor related to the number of entanglements per molecule. The simplest approximation for Q_e can be derived from the dependence of viscosity on molar mass. Experiments with melts of polymers having large molar masses lead to the following correlation (Fig. 58):

$$\eta \sim M^{3,4} \tag{150}$$

so that for the hindrance factor Q_e can be written:

$$Q_e \sim (M/M_c)^{2,4} \tag{151}$$

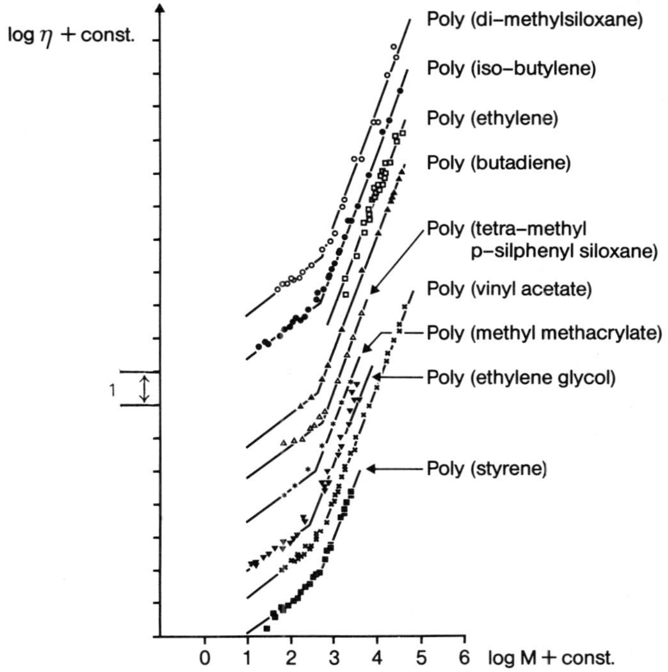

Fig. 58. Molar mass dependence of the viscosity for several polymers (from [52])

Here M_c represents the limiting molar mass at which entanglements start to become relevant. For polymers having broad molar mass distributions the average molar mass M_w should be used. The relaxation-time spectrum and the moduli functions of the Rouse theory remain unchanged when the quantity $Q_e\zeta_0$ instead of ζ_0 is used. Stable physical crosslinks or covalent bonds between polymer molecules lead to rubber elastic deformation behavior. The corresponding modulus plateau can also be described in terms of the Rouse theory by introducing an additional modulus contribution $\varrho RT/M_c$ (N is the number of network chains per unit volume) at infinitely long relaxation times:

$$G'(\omega) = \varrho RT/M_c \left[1 + \sum_{p=1}^{N} \omega^2\tau_p^2/(1+\omega^2\tau_p^2)\right] \tag{152a}$$

$$G''(\omega) = \varrho RT/M_c \sum_{p=1}^{N} \omega\tau_p/(1+\omega^2\tau_p^2) \tag{152b}$$

$$G(t) = \varrho RT/M_c \left(1 + \sum_{p=1}^{N} e^{-t/\tau_p}\right) \tag{153}$$

6.6 Relaxation Processes According to the Meander Model [45, 46]

In the following section the more important relaxation processes and the mechanisms responsible for them in terms of the meander model will be summarized.

6.6.1 Rubber Elasticity

The deformation process responsible for rubber elasticity in the region of the compliance plateau (see Fig. 47) is ascribed, in terms of the meander model, to intrameander shearing (see Fig. 59) whereby layers of molecules in the meander cube are allowed to be displaced by one chain distance d with respect to one another. The resulting plateau compliance, after Reuss-averaging, for an uncrosslinked polymer melt can be written in terms of:

$$J^0_{eN} = \frac{d(r+x)^2}{9kT} \tag{154}$$

where: d is the lateral chain distance, r the diameter of a molecular bundle and $(r+x)$ the height of a meander cube.

The intrameander shear, i.e., shear fluctuations – restricted to one chain distance per layer of molecules – can be used to explain the plateau compliance

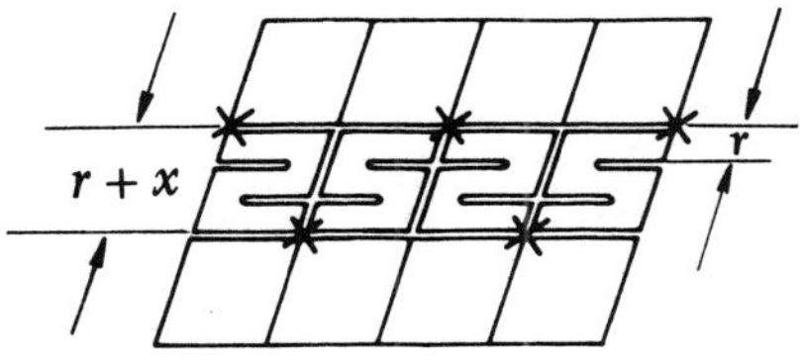

Fig. 59. Two-dimensional sketch of intrameander shear (from [45])

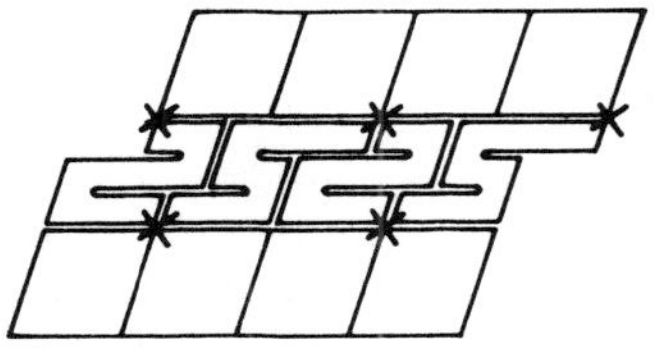

Fig. 60. Shearband deformation (from [45])

at equilibrium. However, they account only for a maximum shear $\gamma_{max}^{intra} \approx 1/\sqrt{3}$. With greater fluctuations large scale deformations become possible by interbundle displacement i.e. by unfolding of suitably arranged meander cubes (see Fig. 60). To this end, whole files of meander cubes or even layers of these must cooperate in a socalled shearband process. The maximum shear deformation by such a process is $\gamma_{max}^{inter} = 9$. For a stress-dependent shearband concentration β the macroscopic relaxation strength of the shearband process (without any loss of orientation within the shear bands) is given by:

$$\Delta J_B^{\infty} = \frac{\beta sd(r + x)^2 \gamma_m^2}{5xkT} \tag{155}$$

(where s is the length of a chain segment) or with a remaining orientation ξ:

$$\Delta J_B = \Delta J_B^{\infty}/\xi \tag{156}$$

If chemical crosslinks are present, they will bridge adjacent layers of molecules within the meander cube. These bridges may hinder the relative displacement of the layers. With the concentration of crosslinks per polymer segment p_c and that fraction which will effectively block the one d-displacements during intermeander shear p_c^* (see Fig. 61), the plateau compliance J_{eN} for a chemically crosslinked polymer can be described by:

$$J_{eN} = J_{eN}^0 \exp\left[-\frac{(r + x)^2}{sd} p_c^*\right] \tag{157}$$

Here J_{eN}^0 corresponds to the plateau compliance for an uncrosslinked polymer (Eq. 154).

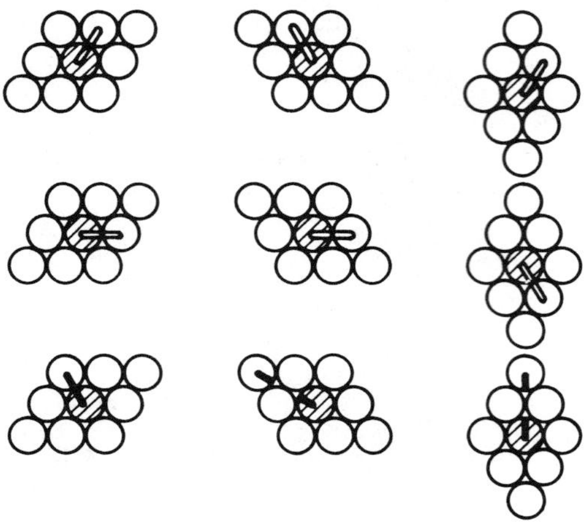

Fig. 61. Ineffective (*open rectangles* in the first two lines) and effective (*filled symbols* in the third line) chemical crosslinks for cross-sectional shear (from [45])

For the relaxation strength of a shearband process which is hindered by all crosslinks present (concentration p_c per segment) Eq. 156 is valid with $\xi \approx p_c/2p_f^0$, where p_f^0 is the fold probability per monomer of length l_0.

Whereas crosslinking strongly changes the form and position of the master curve, a filler content reduces the compliance but keeps its lgω-dependence essentially constant (see Fig. 45 and 46). Quantitatively, the filler effect on the steady

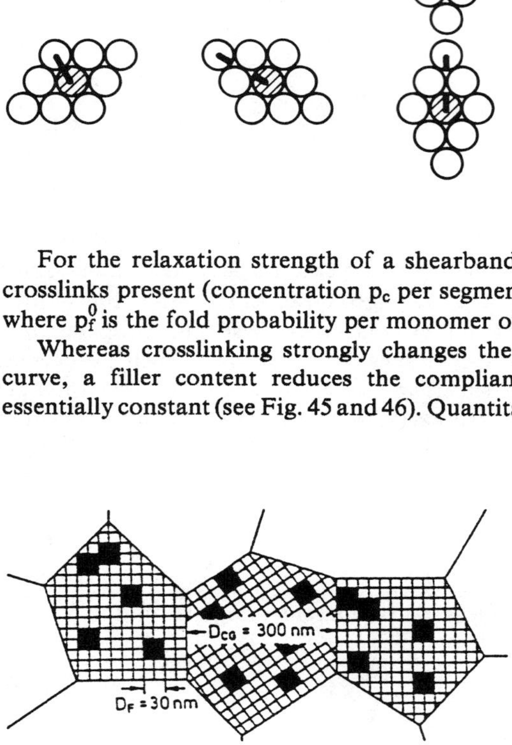

Fig. 62. Coarse grain structure of the polymer matrix with embedded filler particles blocking the shear deformation of several meander files within the grains

state compliance is described on the level of the meander model by its coarse grain structure: the paraelasticity of a file of meanders across a coarse grain of diameter D_{CG} is blocked if it contains at least one filler particle (of diameter D_F). Therefore the reduced compliance is given by:

$$\frac{J(\phi_F)}{J(0)} = W(0) = (1 - \phi_F)^{D_{CG}/D_F} \tag{158}$$

the probability of finding no particle in a file. In this equation ϕ_F is the total volume fraction of filler and D_F and D_{CG} are the geometric quantities shown in Fig. 62.

6.6.2 Flow

The flow relaxation process (see Sect. 6.1) is described by the meander model as a paraelastic deformation attributed to an extra shear (see Fig. 63) caused by the reptating chains, when leaving one junction and entering a neighbouring one. The power law-approximation for the molar mass dependence of the relaxation strength of this process is given by:

$$\Delta J_F^0 \sim (M/M_0)^{0,5} \qquad (159)$$

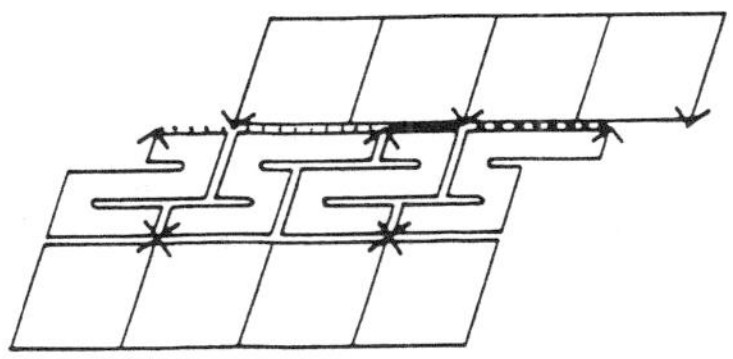

Fig. 63. 2-dimensional sketch of the paraelastic deformation on the upper interface of a shearband, caused by spread junctions (from [45])

The characteristic relaxation time for flow can be written in terms of:

$$\tau_F = \frac{l_0}{4\lambda\Gamma_0}\left(\frac{M_c}{M_0}\right)^3 \cdot \begin{cases} (M/M_c)^3 & \text{for } M \leq M_c \\ (M/M_c)^4 & \text{for } M > M_c \end{cases} \qquad (160)$$

where: λ is the length of the diffusing chain segment; l_0 the length of a monomer unit; M_0 the mass of a monomer unit and Γ_0 is the jump frequency of a monomer unit. $M_c = M_0/p_f^0$ signifies a critical molecular weight, which is close to the M_c-data from viscosity measurements.

The molar mass dependence of the steady state viscosity can be derived from Eqs. 159 and 160:

$$\eta_0 = 2\tau_F/\Delta J_F^0 \qquad (161)$$

6.6.3 Glass Transition Process

In order to explain the glass transition process in terms of the meander model, dislocations have been postulated (see Fig. 64). The concept of chain parallelism, realized by tight folding of molecules into bundles which, in turn, are superfolded to form meander cubes, offers the possibility of introducing dislocations also into an equilibrium melt. Whereas viscous flow in an entangled system can only take place by chain reptation, paraelastic shear fluctuations (i.e., rubber elasticity) are realized by dislocation movements; the latter also being responsible for plastic flow and the anelasticity of crystalline solids. The glass relaxation process corresponds to the freezing of segment mobility when the number of "dislocation segment lines" or

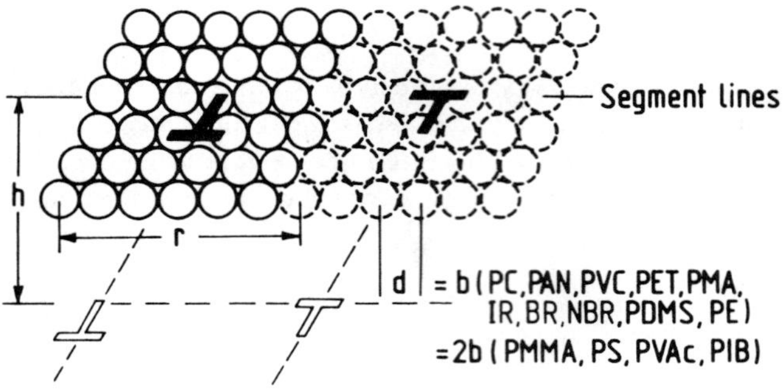

Fig. 64. Cross-section of parallel chains with two edge-dislocation walls of opposite sign. Here: *b* is the Burgers-vector, *d* the chain distance and *h* the spacing of the dislocation array (from [45])

"s-dislocations" (the number of chain segments in one line connecting the two edge-dislocation walls, Fig. 64) having an energy ε_s, decrease below a charcteristic level.

The relaxation frequency of a chain segment is determined by two factors:

(1) by an Arrhenius-type factor accounting for the intramolecular part of the activation and

(2) by a factor arising from molecular interactions and representing the probability that an intramolecularly activated segment will indeed jump. For the intrameander

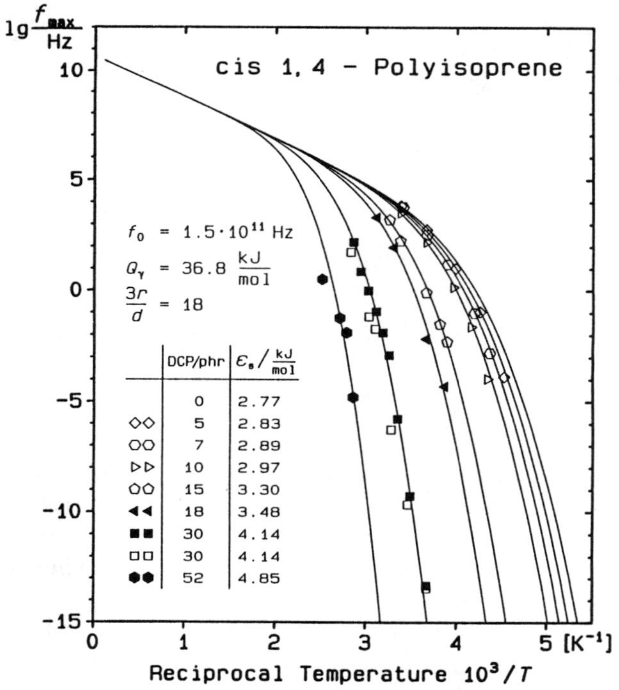

Fig. 65. Activation diagram of polyisoprene crosslinked by dicumyl peroxide (from [45])

shear fluctuations to be fully activated, it is assumed that every segment line across coupled cubes contains at least one segment which is part of a dislocation wall and therefore has an increased energy ε_s. If $\exp(-\varepsilon_s/kT)$ is the probability of finding such a segment at a fixed place, the probability of having at least one of this kind of segment within 3r/d chain segments becomes $1-[1-\exp(-\varepsilon_s/kT)]^{3r/d}$. For fully activated coupled cubes some $3(3r/d)^2 d/s$ segment lines must simultaneously be in this defect state. Putting both factors together the relaxation frequency becomes:

$$f_m = (f_0/\pi)\, e^{-\frac{Q}{kT}} \left[1 - \left(1 - e^{-\frac{\varepsilon_s}{kT}}\right)^{\frac{3r}{d}}\right]^{3\left(\frac{3r}{d}\right)^2 \frac{d}{s}} \tag{162}$$

In Fig. 65 the activation curves for peroxide crosslinked polyisoprene are drawn according to formula (162), varying only the parameter ε_s. The fit to the experimental shift factors derived from the data also shown in Fig. 45 is excellent.

The glass process and shearband relaxation are accelerated by swelling; equivalent to a shift to lower temperatures on a temperature scale. Additionally, the relaxation strength of these processes increases. Putting ϕ equal to the volume fraction of uncrosslinked polymer in the swollen system, the reduced shear compliance for uncrosslinked melts is given by a ϕ^{-2} law:

$$\Delta J(\phi)/\Delta J(1) = \phi^{-2} \tag{163}$$

The corresponding relationship for medium and highly crosslinked polymers is:

$$\Delta J(\phi)/\Delta J(1) = \phi^{-2/3} \tag{164}$$

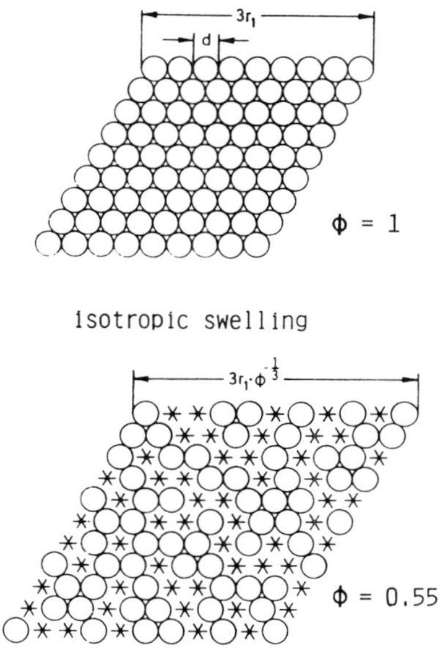

Fig. 66. Isotropic swelling of a meander cube; (O): polymer chains, (∗): solvent molecules (from [46])

This short summary demonstrates that the meander theory often allows a more accurate account of experimental results than can be made using the coil model. The interested reader can find a more detailed discussion of the meander model in the quoted literature from W. Pechhold and coworkers.

6.7 Non-Newtonian Viscosity and the Behavior of Polymers During Processing

The processability of elastomers and thermoplastics is predominantly governed by the macrostructure of the molecules (Molar mass, molar mass distribution, degree of branching, etc.). The macrostructure influences the modes of long range motion of the chain molecules; the latter being accounted for by the long relaxation times in the Rouse spectrum, in particular by τ_1 (Eqs. 143 and 160). Thus, the steady-state viscosity η_0 (Eq. 145 and 161) and the steady-state compliance J_e^0 (Eq. 146 and 159) are the relevant quantities in terms of the processability of polymers. In these terms, η_0 is a measure of the energy dissipated during flow and J_e^0 determines the amount of elastically stored energy, which is manifested during flow in a number of different phenomena (e.g., die swell, elastic turbulence, melt fracture, etc.). It should be noted that both of these quantities are very dependent on the rate of shear.

As a rule, the viscosity of a polymer melt sinks as the shear rate increases by several orders of magnitude (Fig. 67). Such behavior is called structural viscosity. The term emphasizes that in a shear gradient the structures in the melt (e.g. physical networks) are destroyed.

The dependence of viscosity on the shear rate can be described with the exponential equation:

$$\tau = \eta \dot{\gamma}^n \tag{165}$$

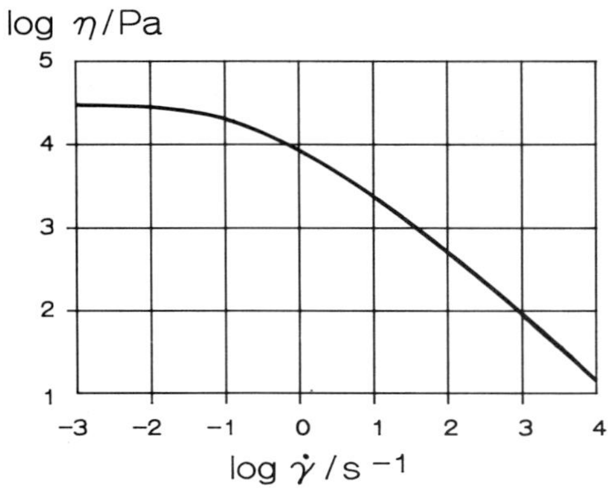

Fig. 67. Schematic diagram showing the dependence of polymer viscosity on shear rate

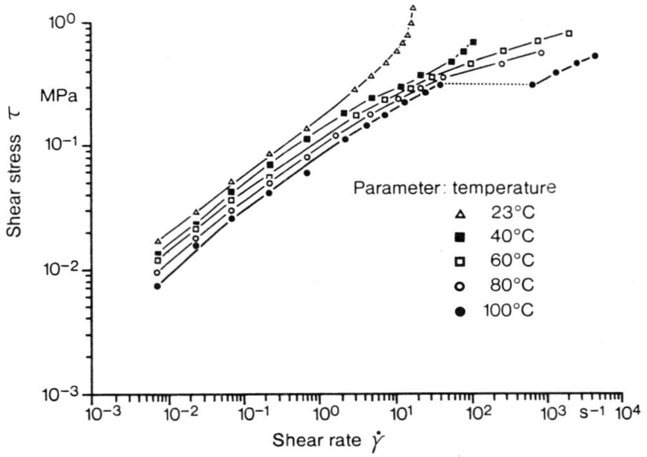

Fig. 68. Flow curves for a stereoregular polybutadiene (98–99% *cis* 1,4 structure) at different temperatures

For the exponent n three cases can be distinguished:

	With hysteresis
a) $n = 1$: Newtonian behaviour	---
b) $n < 1$: Structural viscosity	Thixotropy
c) $n > 1$: Dilatance	Rheopexy

Structurally viscous melts are thixotropic if, at a given shear stress, a decrease in viscosity with time can be observed. In contrast, dilatant melts manifest rheopexy, if the viscosity increases with time under a given shear stress.

Dilatant behavior of a melt means that the apparent viscosity ($\tau/\dot{\gamma}$) increases with increasing shear rate. In this case structures are developed under the influence of a shear gradient. Dilatant flow behavior is often observed for melts of stereoregular polymers with large molar masses, which tend to crystallize under stress or strain. Figure 68 shows flow curves for a stereoregular polybutadiene, synthesized using a uranium catalyst and having a very high degree of stereoregularity. A reduction in temperature leads to a change from structural viscosity to dilatance.

The large-scale manufacturing processes for polymers usually involve considerable shear rates. Figure 69 gives a summary of the more important processes and the associated shear rates. At these large shear rates the behavior of polymer melts is largely determined by their elastic properties.

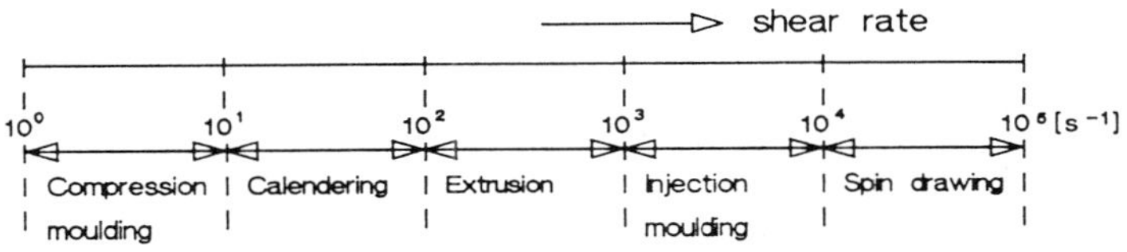

Fig. 69. Shear rates operating during the more important polymer manufacturing processes

The elasticity of a polymer melt leads, under shear flow, to a build-up in the normal stresses, which can be orders of magnitude greater than the shear stresses. The normal stress coefficient, in analogy to the viscosity, can be defined by:

$$\Theta = (\sigma_1 - \sigma_2)/\dot{\gamma}^2 \tag{166}$$

It varies qualitatively in the same way as the viscosity (see Fig. 67). Additional phenomena whose origins lie in the elasticity of the polymer melt are, for example, die swell and melt fracture. In particular, melt fracture can lead to considerable difficulties, such as the development of extraneous structures in the melt, during polymer processing. Figure 70 shows the flow curves for a polybutadiene, synthesized using a Li-catalyst, which show irregularities suggesting the presence of melt fracture.

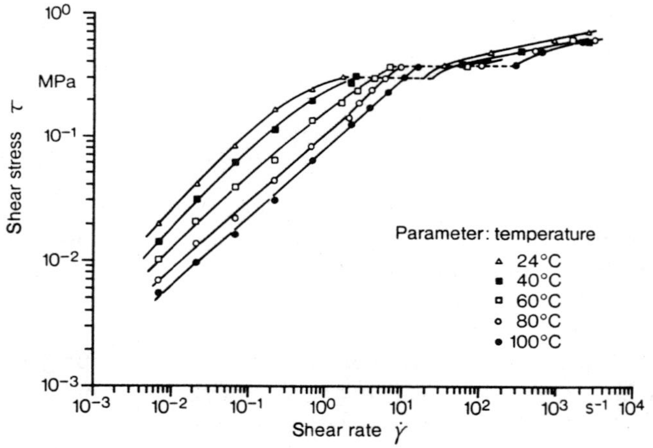

Fig. 70. Flow curves for Li-polybutadiene

The onset of melt fracture is manifested by a sudden increase in extruded volume from a capillary viscosimeter with only a small increase in extrusion pressure. This effect is indicative of a reduction in adhesion between the polymer melt and the wall of the die and the initiation of slip processes. Generally, melt fracture leads to pronounced irregularity of the extrudate surface and in extreme cases to breaking of the extruded strand. In Fig. 71 the effects of melt fracture on the appearance of rough sheets on a two roll mill are shown.

The sheets tend to tear and form pouches due to a loss of elasticity in the sheet. This loss of elasticity from the rubber melt can be attributed, in this case, to a breakdown of the physical network after a critical shear rate is exceeded. Quantitative estimates of the critical parameters lead to the conclusion that melt fracture occurs when the number of physical crosslinks has decreased to about 1/3 of the original number in the unstressed polymer.

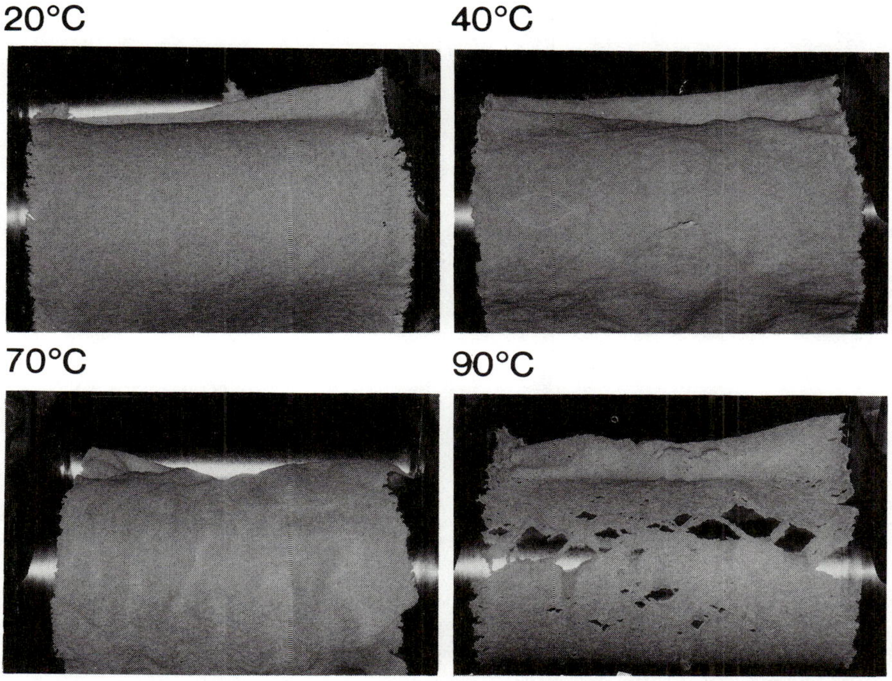

Fig. 71. Appearance of rough sheets of Li-polybutadiene on a two-roll mill at various temperatures

Part II
Crystallization and Melting of Polymers

7 Crystallization Behavior [58–75]

7.1 Polymer Crystals and Growth Forms

Single polymer crystals were first recognized in 1953 (Schlesinger and Leeper). Such crystals grow preferably from dilute solution to form rhombic platelets several μm in diameter and approx. 10 nm thick.

Figure 72 shows single polyethylene crystals, grown from a dilute solution, in which spiral, layered structures can be interpreted. Since the layers have a maximum thickness of several dozen nm and the polymer chains are several hundred nm long and orientated normal to the layer surfaces, it is obvious that chain folding must take place during crystallization. An acceptance of the principle of chain folding for which up to now, no satisfactory, thermodynamic theory exists, requires a modification of the Fransen micelle model for partially crystalline polymer solids (Fig. 73).

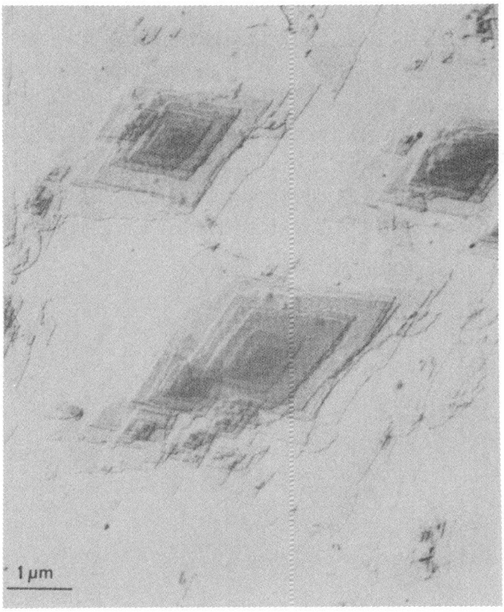

Fig. 72. Electron micrograph of single PE crystals (from [58])

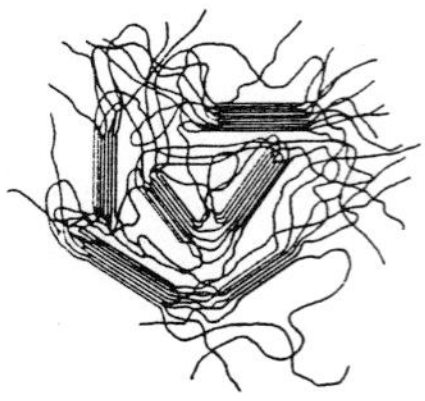

Fig. 73. Fransen micelle model for partially crystalline polymer solids

According to Wunderlich there are three possible macroconformations for the molecules in a polymeric solid (see Fig. 74).

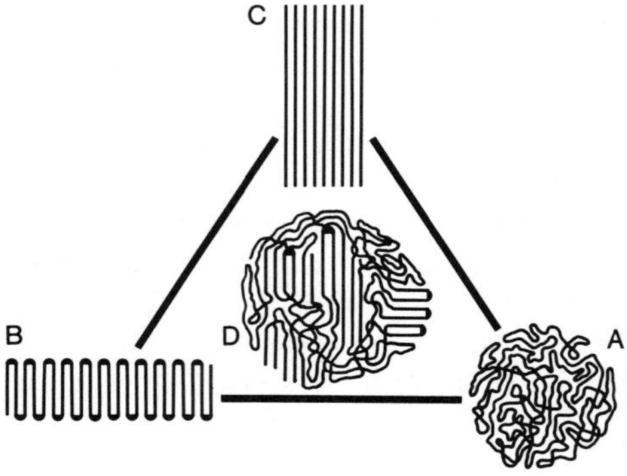

Fig. 74. Sketch showing the three possible macroconformations for the molecules in a polymeric solid (from [59])

The morphology of a partially crystalline polymer is usually composed of all three types of macroconformation. For the chain folds in a single lamella various models can be put forward (see Fig. 75).

When crystallization occurs from a melt under increased pressure, the individual fold-lamellae become thicker and the chains approach a stretched form to give extended chain crystals. For polyethylene it has been shown that the lamella-thickness also increases with increasing temperature. Lamella-crystals are obvious-

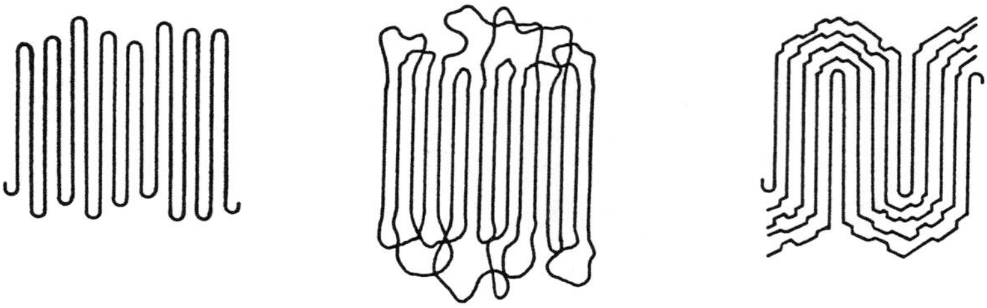

Fig. 75. Models of crystal-lamellae
a. Acute folds, adjacent reentry; b. loose folding; c. meander model according to Pechhold

ly not an equilibrium form but rather an expression of a metastable state. An extended chain crystal is closer to a thermodynamic equilibrium state than a folded-crystal.

The nature of the supramolecular structure in a crystal from a polymer melt is dependent on the conditions prevailing during crystallization; in particular, the nucleus distribution. A common form is radially symmetrical spherulites which develop during crystallization in three dimensions and whose growth is inhibited only by their coming into contact with an adjacent spherulite. Figure 76 shows, schematically, the growth of such a spherulite. It consists, in its middle, of radially ordered crystal lamellae in which the polymer chains are, in turn, ordered normal to the direction of growth. Figure 77 shows the ordering of the lamellae and the polymer chains in the area around the core of a spherulite. Between the crystal-lamellae the polymer is amorphous.

If there is a high concentration of nuclei then fine crystal structures develop and the material will be comparatively transparent. On the other hand, if only a small number of nuclei are present, coarser structures develop and the material will be opaque. These effects can be very important technologically, for example, during

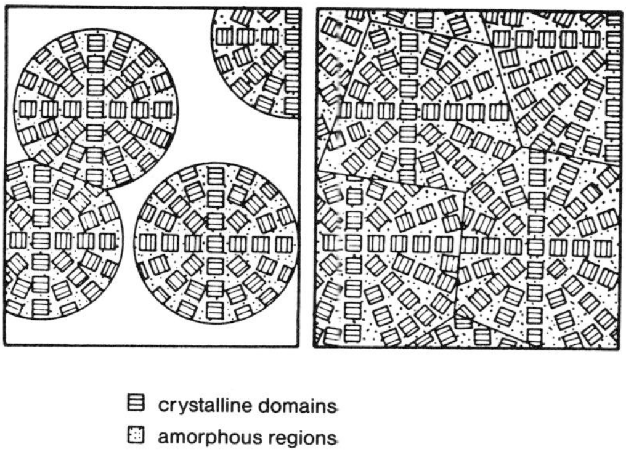

Fig. 76. Sketch showing the ordering of the lamellae in a spherulite (From [60])

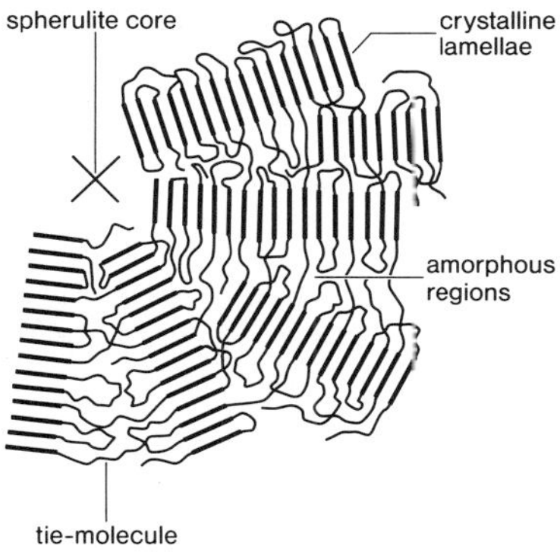

Fig. 77. Ordering of the lamellae in the region around the core of a spherulite. Note: the picture is not drawn to scale, thus the ratio of the spherulite radius to the thickness of the lamellae is not correct

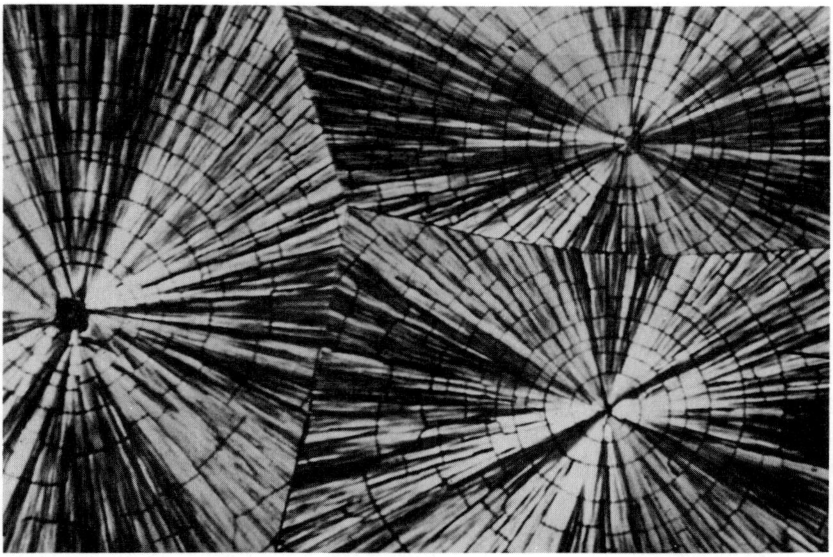

Fig. 78. Spherulites in polypivalolactone

the manufacture of thin films. Figure 78 shows the spherulites in a polypivalolactone photographed in polarized light.

The socalled tie-molecules are of critical importance for the technological properties of partially crystalline polymeric materials. These tie molecules form bridges between two or more lamellae and are responsible for the cohesion between lamellae. They are also responsible for the transfer of stress through the material.

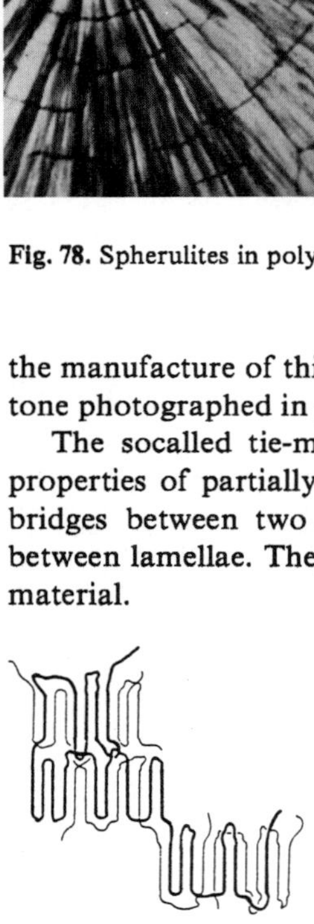

Fig. 79. Schematic representation of a single macromolecule transversing different lamellar crystals. Clusters of crystalline stems situated in different lamellae are connected by tie molecules. Note that the drawing is not in correct scale: The length of the crystalline stems is about 50 times larger than their lateral distances (from [68])

Neutron scattering experiments by E. W. Fischer corroborate assumptions according to which the well-defined clusters of crystalline stems in a single lamella belong to a single molecule (Fig. 79). Such clusters would not be formed if a single molecule were responsible for a large number of tie-sequences between two adjacent lamellae. Fischer explains the existence of well defined clusters by suggesting that the growth fronts within the lamellae of a stack do not arrive at the position of a single macromolecule simultaneously. The situation is shown schematically in

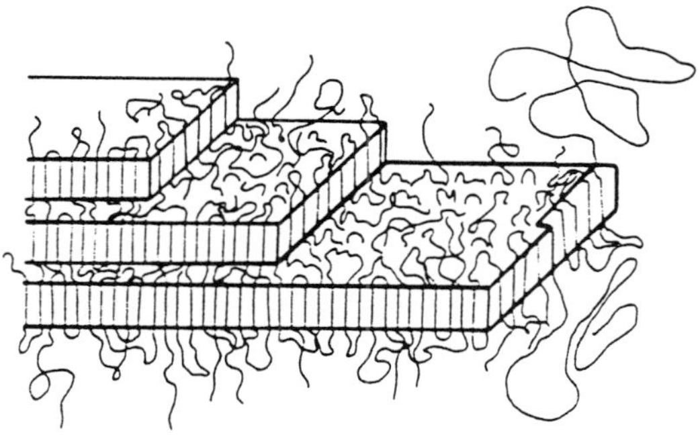

Fig. 80. Schematic diagram of the growth of a stack of lamellae in the melt. The growth fronts do not arrive simultaneously at the location of a single molecule (from [68])

Fig. 80. The stepwise growth will result in approximately ($\nu - 1$) tie-molecules if ν is the average number of clusters per molecule. The growth of the cluster within a single lamella is, according to Fischer's model, halted by kinetic hindrance; e.g. by entanglements and by a filling in of the growth front with parts of other molecules.

An electron micrograph of a polyethylene/paraffin mixture after the paraffin has been removed is shown in Fig. 81. Such micrographs can be interpreted to corroborate the existence of tie-molecules in interlamellar bridges.

It is known that breaks in partially crystalline polymers occur preferentially along spherulite boundaries or within the spherulites along the lamellae boundaries.

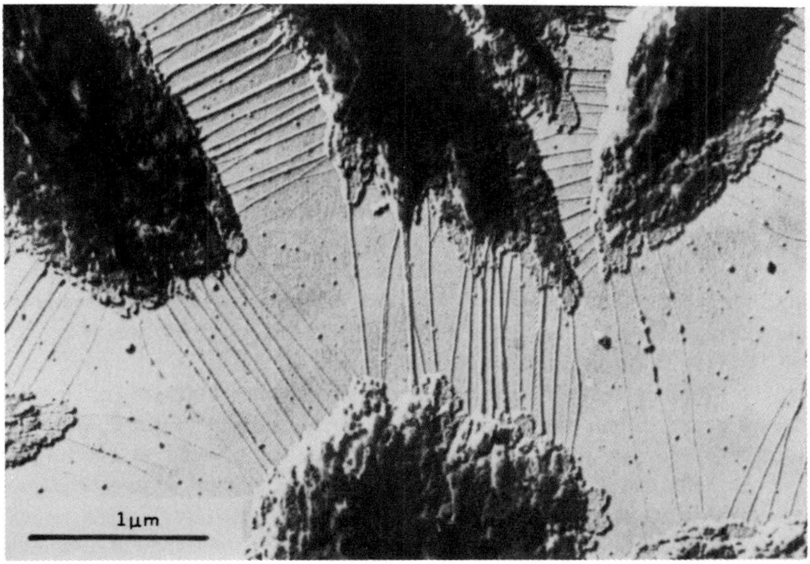

Fig. 81. Interlamellar bridges in polyethylene ([from 72])

Thus, the toughness and tensile strength of partially crystalline thermoplastics are directly proportional to the concentration of the intercrystalline segments in the amorphous regions.

7.2 Crystalline Structures in Stretched Polymers

A crystalline structure can also be induced in polymer melts or solutions by stress or strain. Using this technique Andrews, in 1966, was able to observe filament structures in highly elongated natural rubber films (Fig.82).

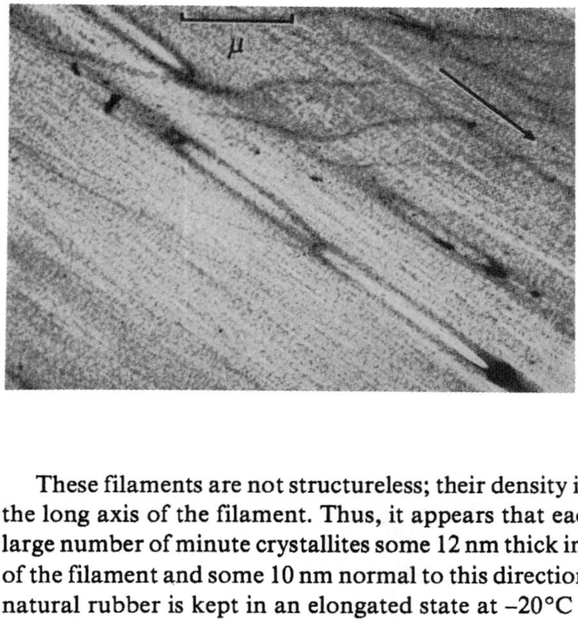

Fig. 82. Electron micrograph of filament structures in a highly elongated (approx. 700%) natural rubber film (from [73])

These filaments are not structureless; their density increases and decreases along the long axis of the filament. Thus, it appears that each filament is composed of a large number of minute crystallites some 12 nm thick in the direction of the long axis of the filament and some 10 nm normal to this direction. If such a vulcanized film of natural rubber is kept in an elongated state at –20°C it will continue to crystallize normal to the length of the filaments (Fig. 83) to form, socalled, "shish-kebab" structures. The latter are shown schematically in Fig. 84.

Pennings and Keller treated such filaments for some 80 h with fuming nitric acid. Since, even after this drastic treatment, a fibrillar nucleus could still be observed, it was assumed that the crystals were of an extended chain type. According to Pennings, the nucleus of a filament is composed of extended chain crystallites onto which, at regular distances, lamella crystallites grow epitactically (Fig. 84). More recently, it has been suggested that the fibrils are produced by the growth of lamella crystallites in a spiral form. Generally, such stress induced crystals melt on removal of the external force.

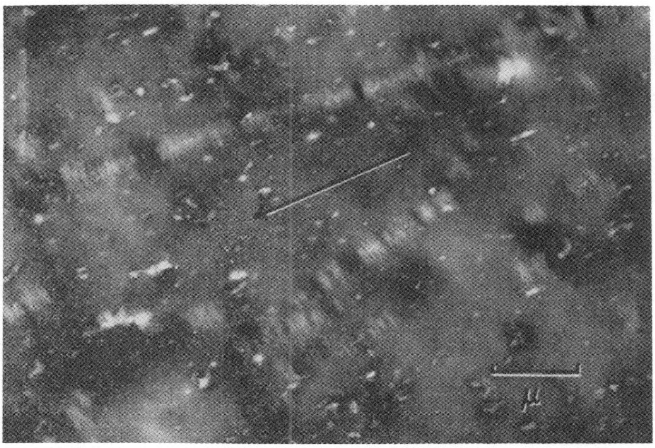

Fig. 83. Film of natural rubber after being held in an elongated state at –26°C for 1 h (from [73])

Fig. 84. Sketch of a "shish-kebab" filament (from [74])

A rather different situation exists if, by stretching partially crystalline materials, an orientation is introduced and then frozen in. The nature of the colloid structure produced in this way is strongly dependent on the temperature at which the material is stretched. At higher temperatures the process operates homogeneously over the whole sample whereas, at temperatures below the melting point of the sample, the stretching operates only in isolated regions of the sample. The flow processes in these small regions lead to necking, i.e. the stretching is inhomogeneous.

Stretching (or drawing) and orientating are technologically important processes. In particular, in the manufacture of fibers, which can be drawn from the melts or

Fig. 85. Colloid structure of a stretched fiber

concentrated solutions of most polymers. Such filaments are usually not suitable for making textiles since, as formed, they lack the required strength. Normally, the initially formed filaments are stretched after exiting the spin-die. This stretching leads to an irreversible elongation of the filament. During this process the amorphous polymer chains are uncoiled and orientated in the direction of the fiber-axis. Additionally, the lamellae structures are partially unfolded as the remaining lamellae turn to become orientated in the direction of the fiber-axis. Figure 85 is a sketch of such a drawn fiber. Here too, the tie molecules, their number and degree of orientation, are critical for the strength of the fiber.

7.3 Nucleation

There are two principal types of nucleation: thermal and athermal. The origins of thermal nucleation are fluctuations, produced by random molecular motions which lead to the formation of crystal-like regions; socalled embryos. Such regions are unstable above the melting point of the material. Below the melting point there is a temperature-dependent, critical nucleus size. Embryos which are larger than this critical size are capable of growth and can thus become actual crystal nuclei. Überreiter has proposed the following equation for describing the rate of nucleus formation:

$$\dot{N} = \dot{N}_0 \frac{T}{\eta(T)} e^{-\Delta G_{max}/RT} \tag{167}$$

where ΔG_{max} is the energy of activation for nucleus formation, $\eta(T)$ is the viscosity of the material and $\dot{N}_0$ is a normalizing factor.

Initially, the rate of nucleus formation increases as the difference between the material temperature and its melting point increases due to a decrease in the activation energy of the process (see Fig. 86). With further temperature decrease the rate of nucleus formation goes through a maximum and decreases rapidly as the material approaches its glass transition temperature.

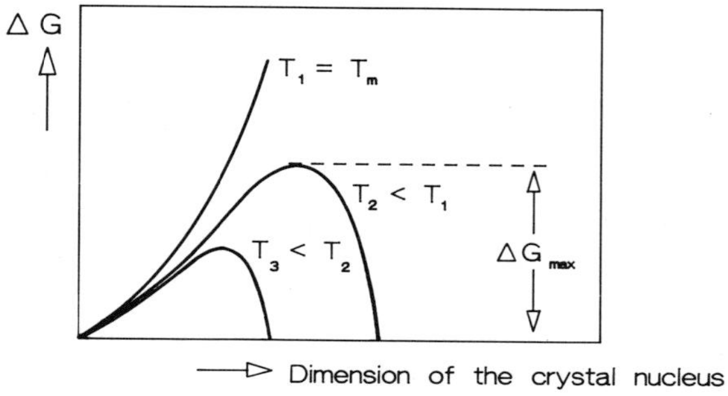

Fig. 86. Free energy of crystal nuclei formation as a function of temperature

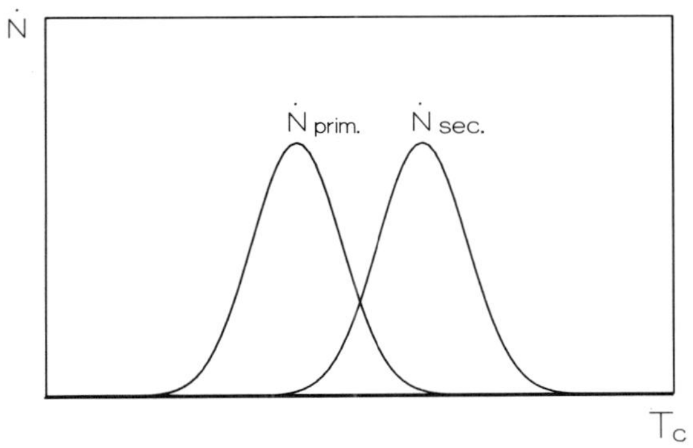

Fig. 87. Primary and secondary nucleus formation rates as a function of crystallization temperature T_c

In Eq. 167 the diffusion term $(T/\eta(T)$ is responsible for the decrease in the rate of nucleus formation at lower temperatures. The overall rate of nucleus formation as a function of temperature is shown schematically in Fig. 87.

An athermal nucleus formation is the process whereby, due to a rapid drop in temperature, a crystal embryo becomes a stable nucleus. In order to be stable at a given temperature the embryo must exceed the critical size already discussed.

It is also possible for extraneous particles to form crystal nuclei if they fulfill the following conditions: They must

a) reduce the boundary surface energy
b) have a melting point above that of the polymer
c) be insoluble in the polymer melt
d) be involatile and
e) be chemically inert with respect to the polymer.

Provided these conditions are fulfilled, a heterogeneous nucleus formation can occur.

7.4 Crystal Growth [75]

The growth of crystals occurs due to secondary and tertiary nucleation. Since the new interfacial area required by nucleation of higher order is smaller, the activation energy for these processes is reduced. Thus, the maximum rate of secondary nucleation occurs at higher temperatures than the primary nucleation (see Fig. 87). This leads to the number and size of the growing crystallites being dependent on the crystallization temperature T_c. At a constant degree of crystallization, crystallization at higher temperatures leads to a smaller number of larger crystals whereas if considerably supercooled, a large number of small crystals form. As crystallization progresses diffusion (Eq. 167) plays an ever increasing role, since chain-segment transport over ever increasing distances is required. Thus, for example, crosslinking may lead to an increase in the rate of primary nucleation due to a more favorable orientation of neighboring chain segments. However, crosslinking reduces the rate of crystal growth due to it hindering the diffusion of individual chain segments within the matrix.

The course of crystallization can be followed dilatometrically; the decrease in volume associated with crystallization is measured as a function of time. A useful measure of the rate of crystallization is the half-life $t_{1/2}$, the time taken for half the crystallizable material to crystallize. The dependence of the half-life on the crystallization temperature is parabolic since the rate of crystallization tends to zero at both the equilibrium melting temperature and the glass transition temperature (Eq. 167). In Fig. 88 the half-lives for the crystallization of several polymers as a function of crystallization temperature are plotted. It can be seen that polyisoprene crystallizes considerably slower than a polybutadiene with a similar *cis/trans* composition. This can be attributed to the crystallization inhibiting effect of the methyl side groups. On the other hand, natural rubber, which is essentially 100% *cis*-1.4-polyisoprene, crystallizes faster than a synthetic polyisoprene with only 2% *trans*-1.4 sequences. Figure 88 also makes clear that polymers such as polypentenamer and polychloroprene with a predominant trans-configuration crystallize faster than their predominantly *cis* counterparts.

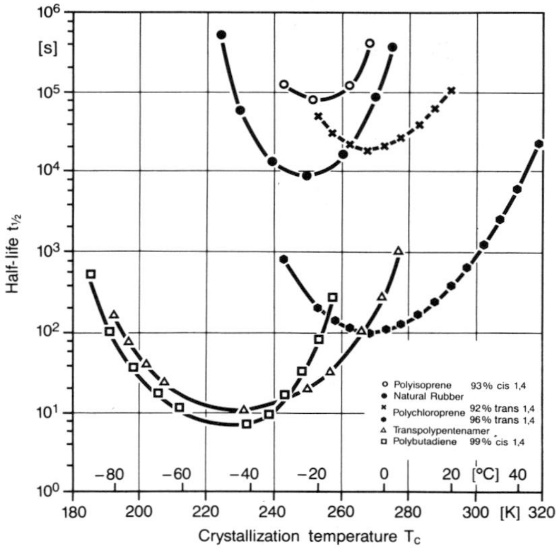

Fig. 88. Crystallization half-lives for several polymers as a function of crystallization temperature

In Fig. 89 the crystallization half-lives of crosslinked polymers are plotted as function of temperature. A comparison with Fig. 88 shows that the crystallization of all elastomers is more or less retarded after crosslinking. The amount of retardation for natural rubber and polyisoprene is so pronounced that the time axis of the figure is too small.

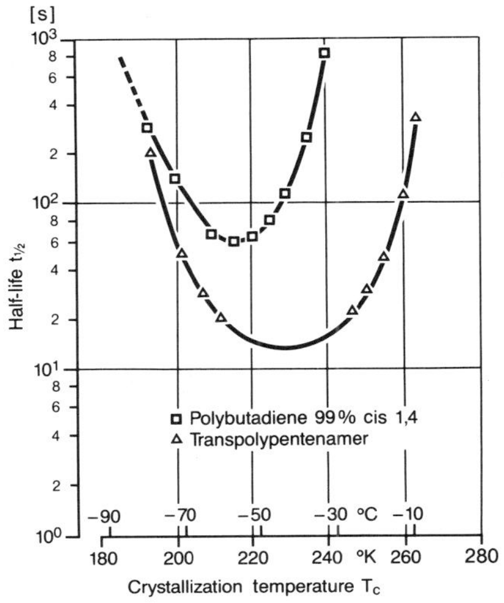

Fig. 89. Crystallization half-lives for cross-linked elastomers as a function of crystallization temperature

In many cases, it is possible to describe the time dependence of the degree of crystallization for an isothermal crystallization with the socalled Avrami equation:

$$\alpha(t) = 1 - e^{-kt^n} \tag{168}$$

In this equation n is the Avrami exponent and k is a quantity which is proportional to the number of nuclei per unit volume and the rate of nucleus formation; it is dependent on the geometry of the growing crystals. The Avrami exponent is always a whole number and provides information about the nature of the crystal growth (see Table 6).

Table 6. Avrami exponent and the dimensionality of crystal growth

Avrami exponent	Type of growth with	
	constant nucleus concentration	constant rate of nucleus formation
1	one-dimensional	–
2	two-dimensional	one-dimensional
3	three-dimensional	two-dimensional
4	–	three-dimensional

For polymers which do not completely crystallize a modified form of the Avrami equation is often used.

$$\alpha(t) = \alpha_0(1 - e^{-kt^n}) \tag{169}$$

In this equation α_0 is the maximum possible degree of crystallization. For crystallizations which can be described by the Avrami equation the appropriate (see Eq. 170) plot yields a straight line whose slope is the Avrami exponent n:

$$\ln\left[-\ln\left(\frac{\alpha_0-\alpha}{\alpha_0}\right)\right]=\ln k+n\ln t \tag{170}$$

By such, socalled, Avrami plots it is not uncommon for considerable deviations from a straight line to result and also for the slope to deviate from a whole number. One reason for such deviations is the phenomenon of post-crystallization. This process leads to an increase in the degree of crystallization due, for example, to conformational changes in the sufaces of the crystal lamellae.

If a polymer, which has crystallized after considerable supercooling, is slowly warmed, a recrystallization can occur. During this process the originally, small and distorted crystals melt and recrystallize to larger crystals with higher melting points, so that the melting temperature range of the material becomes broader.

8 Melting Behavior [76–84]

8.1 Equilibrium Thermodynamics

The melting of polymers is, in principle, a first order transition. Nevertheless, it is not possible to describe all the experimental observations, which have been made for polymer melts, in terms of equilibrium thermodynamics. Due to the limited mobility of the long polymer chains, they do not reach their equilibrium conformation within a finite time. Thus, in order to completely describe the state of a system, not only the usual variables of state but also inner ordering parameters, which reflect the thermal history of the system, are required.

The equilibrium state of an isothermal, isobaric system is characterized by a minimum in the Gibbs free energy:

$$G = H - TS \rightarrow \text{minimum} \quad \text{or} \tag{171}$$

$$dG = dH - TdS = 0 \tag{172}$$

Such a state can be reached via two paths: by minimizing the internal energy (or the enthalpy, H) or by maximizing the entropy, S. In the solid state of a crystallizable material the tendency to minimize the energy predominates whereas in the liquid state, the desire to maximize the entropy predominates. In a plot of $G = f(T)$ the slope for the melt, starting from a higher temperature, is greater than that for the crystal (see Fig. 90).

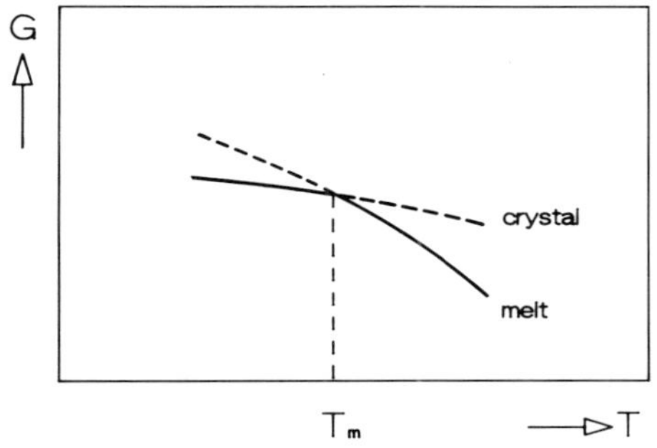

Fig. 90. Sketch of the Gibbs free energy as a function of temperature for a crystallizable substance

From the equilibrium conditions (Eq. 171) it follows that below the point of intersection (Fig. 90) only the crystal state is stable; at temperatures higher than the point of intersection, only the melt is stable. The temperature at which both phases are stable is called the equilibrium melting point T_m^0. From Eq. 172 it follows that:

$$T_m^0 = \Delta H_0/\Delta S_0 \tag{173}$$

The enthalpy and entropy of melting ΔH_0 and ΔS_0 are defined as the changes in state for the conversion of 1 Mole from the crystal to the melt phase.

As already mentioned, partially crystalline materials are not usually in their thermodynamic equilibrium state. Thus, the usually quoted melting points are rarely the equilibrium values. Hoffman and Weeks have developed a method of extrapolating experimental values to obtain the equilibrium melting point. Thus, the melting points T_m are plotted against the temperature at which the material crystallized. The intersection of an extrapolation of the resulting straight line with the line $T_m = T_c$ gives the equilibrium melting temperature (Fig. 91).

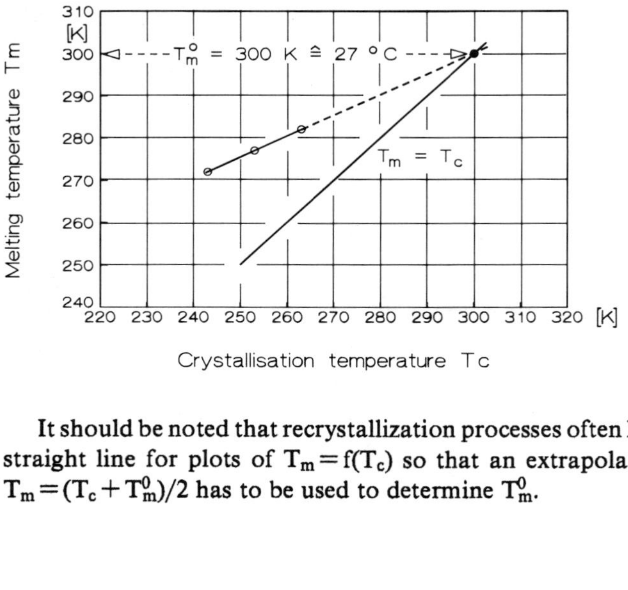

Fig. 91. Melting temperature as a function of crystallization temperature for polyisoprene

It should be noted that recrystallization processes often lead to deviations from a straight line for plots of $T_m = f(T_c)$ so that an extrapolation of the straight line $T_m = (T_c + T_m^0)/2$ has to be used to determine T_m^0.

8.2 Influence of Crystallite Size

The equilibrium melting point corresponds to a transition for phases of infinite size. However, in real systems the phases are of finite size. The individual crystallites have surfaces, whose energies cannot be neglected in terms of the total energy of the considered volume. These surface energies lead to a reduction in the enthalpy of melting and, according to Eq. 173, a corresponding decrease in the equilibrium melting point.

For a cubic crystal with edge length a and a specific surface energy σ:

$$\Delta H_{Gf} = 6a^2\sigma \tag{174}$$

or per Mole:

$$\Delta H_{Gf}/\text{mol} = \frac{6a^2\sigma}{a^3}\frac{M}{\varrho_c} \tag{175}$$

Equation 173 can then be modified to give:

$$T_m = \frac{\Delta H_0 - \Delta H_{Gf}}{\Delta S_0} \tag{176}$$

or with Eq. 175:

$$T_m^a = \frac{\Delta H_0}{\Delta S_0}\left(1 - \frac{6\sigma}{a\Delta H_0}\frac{M}{\varrho_c}\right) \quad \text{or} \tag{177}$$

$$T_m^a = T_m^0\left(1 - \frac{6\sigma}{a\Delta H_0}\frac{M}{\varrho_c}\right) \tag{178}$$

Equation 178, originally derived by J.J. Thompson, indicates that the melting point of a partially crystalline polymer will deviate more strongly from T_m^0 the smaller the crystallites are. As a rule, polymeric materials contain a range of crystallite sizes so that they melt over a broad temperature range.

8.3 Entropy Effects

8.3.1 Stress Crystallization

If the coiled polymer molecules in a polymer melt are stretched, their entropy approaches that of the crystalline state. The difference between the entropies of the orientated and crystalline states decreases and thus, as indicated by Eq. 173, the melting point is raised. If the melting point is raised above that of the environment, stress crystallization takes place. The following relationship was derived by Flory for the dependence of the melting point on strain for a crosslinked elastomer:

$$1/T_m - 1/T_m^0 = -(R/\Delta H_0)g(\lambda) \tag{179}$$

The function $g(\lambda)$ depends on the type of chain statistics applied. According to Flory:

$$g(\lambda) = (6/\pi n)^{1/2}\,\lambda - (\lambda^2/2 + 1/\lambda)/n \tag{180}$$

where n is the number of statistical chain segments per network chain and $\lambda = l/l_0$ is the strain. For highly elongated natural rubber networks, increases in melting point of up to 80 K with respect to the equilibrium melting point ($T_m^0 = 30\,°C$) have been observed.

The following equation has been proposed by Kraus and Gruver for crosslinked trans-polypentenamer:

$$T_m = 290 + 7(\lambda - 1) \tag{181}$$

Stress crystallization plays a crucial role in the strength of elastomeric materials. It is particularly relevant for their tear strength (or cut growth resistance). Thus, when a small tear forms in a rubber article, the external application of force leads to a stress concentration at the base of the tear. Consequently, the tear propagates. Such processes as knicking and tear propagation, are responsible, for example, for the catastrophic groove cracking of truck tires (chipping and chunking) and, microscopically, for abrasion. The tear strength of rubbers is considerably increased by stress crystallization. That this is so, can be demonstrated by Fig. 92.

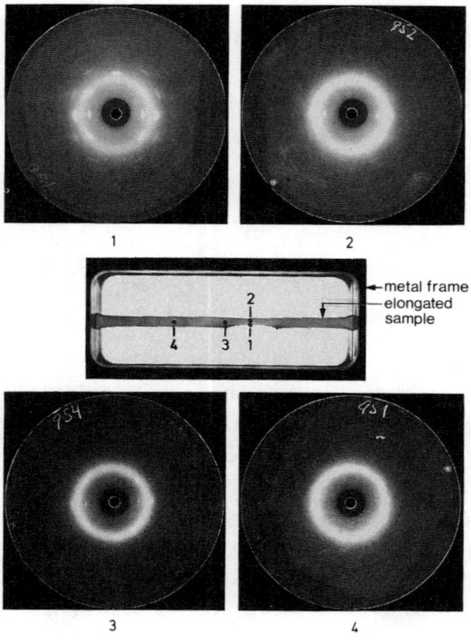

Fig. 92. X-ray diagrams at several positions from an elongated sample ($\lambda = 4$) of unfilled, crosslinked natural rubber, which was cut at position 1 with a razor blade

Here a strip of unfilled, crosslinked natural rubber was cut, extended to three times its original length and clamped so that the deformation remained constant (Fig. 92, middle). The sample was then X-rayed at various points around the cut. From the pictures obtained it is clear that stress crystallization has taken place at the base of the cut (Positon 1) and that as the distance from the cut-base increases the stress crystallization becomes less pronounced (positions 1 to 4). Indeed, no crystallization has occured at position 4. The stress-induced crystallites at the base of a cut reinforce the rubber and block tear propagation; the tear strength is increased.

8.3.2 Entropy of Mixing

Melting point depression is observed for polymer mixtures which are compatible in the melt but which cannot form mixed crystals. Since the separate crystallization of the two components is equivalent to a demixing, the entropy of the melt includes an excess entropy of mixing. For mixtures composed of a crystallizable and a non-crystallizable high polymer the depression of the melting point is given, according to Nishi and Wang, by:

$$1/T_m - 1/T_m^0 = -(R/\Delta h_2)(v_2/v_1)\chi(1 - \phi_2)^2 \qquad (182)$$

where the subscript 1 refers to the non-crystallizable polymer and 2 to the crystallizable one. Furthermore:

v_1, v_2: Molar volumes of the monomer units
Δh_2: Molar enthalpy of melting/monomer unit
ϕ_2: Volume fraction of component 2
χ: interaction parameter

It can be seen that, according to Eq. 182, a melting point depression only occurs if the interaction parameter is negative i.e., when the two components are compatible in the melt.

Part III
Non-linear Deformation Behavior of Polymers

9 Mechanism of Deformation of Thermoplastics and Multi-component Systems [85–97]

9.1 Terminology

The deformation of polymers in the non-linear region is exceedingly complex. Even in the simple, uniaxial stress-strain experiment there is a considerable variety of possible phenomena. Figure 93 shows stress-strain curves for a selection of thermoplastics and elastomers.

The curves are representative of brittle (*Curve 1:* SAN (Styrene-Acrylonitrile)), more or less pronounced ductile (*Curves 2–5:* ABS (Acrylonitrile-Butadiene-Styrene), polycarbonate, polyamide and polypropylene), and rubber elastic (Curves *6* and *7*: NR/BR and PUR) deformation behavior. The deformation characteristics are determined, not only by the chemical nature of the polymer (the microstructure) but also, indeed predominantly, by the supramolecular or stress induced structures. A characteristic of ductile deformation is a yield maximum followed by a flow region. A "yield stress" is indicative of an inhomogeneous stretching of the sample which results in a pronounced necking of the test-piece. Characteristic of elastomers is the reversibility of deformations at low moduli. At greater degrees of strain elastomers tend to self-reinforcement (increasing slope of the stress-strain curve

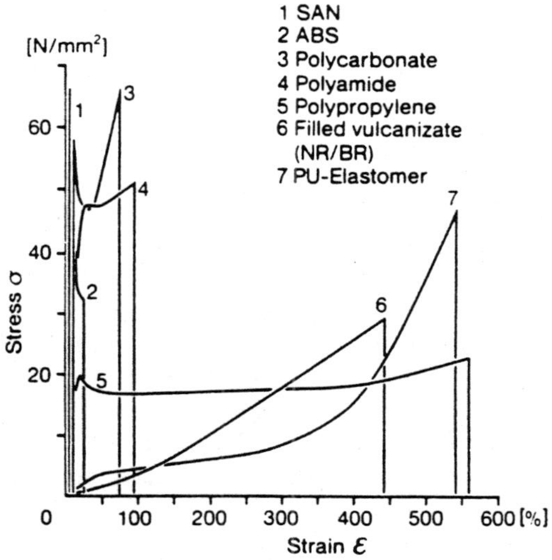

Fig. 93. Stress-strain curves for different types of polymers

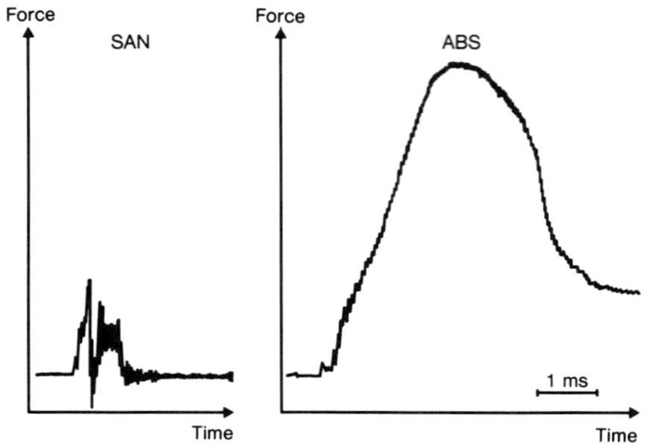

Fig. 94. Force-time curves recorded during a falling dart experiment (*left:* SAN, *right* ABS (from [85])

with increasing elongation) a phenomenon which is generally responsible for the tensile strength of such materials. Self-reinforcement can also be observed for some thermoplastics, for example, polycarbonate. In these cases the resulting toughness of such materials is responsible for their technological importance.

A common method of quantifying the toughness of thermoplastics is the measurement of their impact strength or their notched impact strength using a flexed beam impact tester or an impact pendulum. A disadvantage of these methods is that the results are not material constants but rather a product of e.g. the sample geometry. Another experiment which can be used to quantify the toughness of polymers is the free-falling dart impact test, whereby a dart is dropped from a given height onto a thin plate of known thickness. The force/time curve during the piercing process is then recorded. Examples of such curves for a brittle (SAN) and a ductile (ABS) material are shown in Fig. 94. The integral of the force/time curve with a known dart velocity is a measure of the toughness of the material in terms of the energy absorbed.

9.2 Crazing

From Fig. 94 it can be seen that considerably more energy is required to pierce a sample of ABS than is neeeded for SAN. ABS is a multi-phase system composed essentially of an SAN matrix containing dispersed rubber (polybutadiene) particles (see Fig. 95).

The elastomer particles initiate the deformation mechanism which is responsible for the toughness of ABS. This mechanism does not involve the absorption of energy in the elastomer particles but rather the formation of highly stretched material in zones, known as crazes, around these particles. Figure 96 shows a model of this energy absorbing deformation process, whereby each elastomer particle is a starting-point for a single craze. The rubber particles induce an increase in stress which leads to the development of numerous crazes, more or less simultaneously, in a larger portion of the mechanically strained sample volume (see Fig. 97).

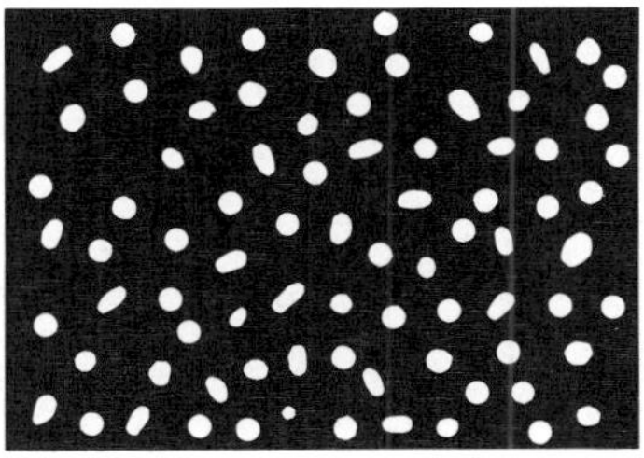

Fig. 95. Schematic representation of the two-phase morphology of ABS

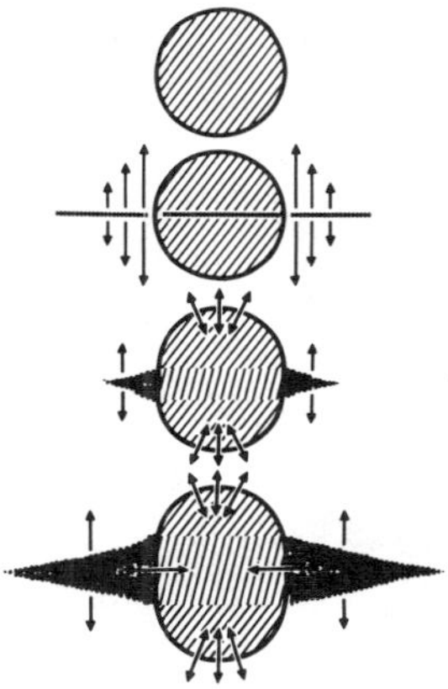

Fig. 96. Model of craze formation (from [85])

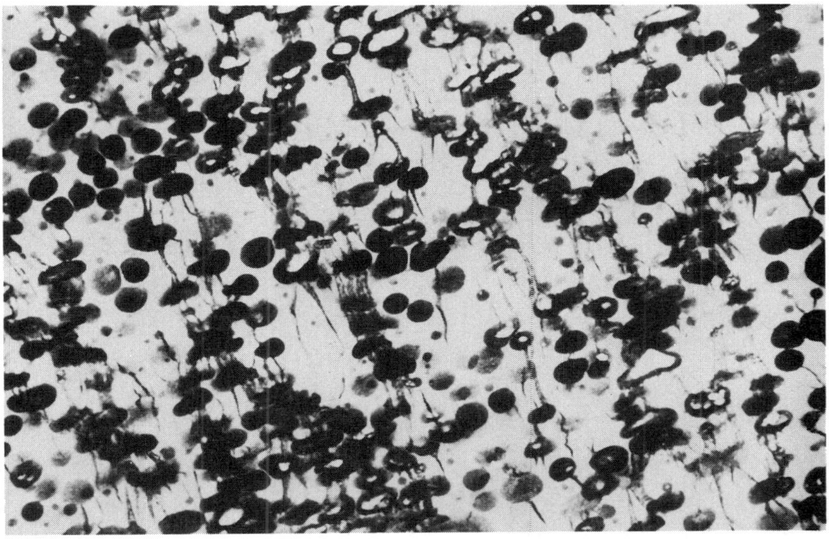

Fig. 97. Electronmicrograph of crazing in ABS (from [85])

Within a craze-band the chains are in a stretched state. Since, however, each individual stretching process requires energy, the external deformation work is divided into a large number of smaller units which are then spread over the whole of the deformation zone.

Figure 98 is an electronmicrograph (TEM) of a single craze-band in polystyrene. In the centre of the band one can see a more open, less dense structure. Craze

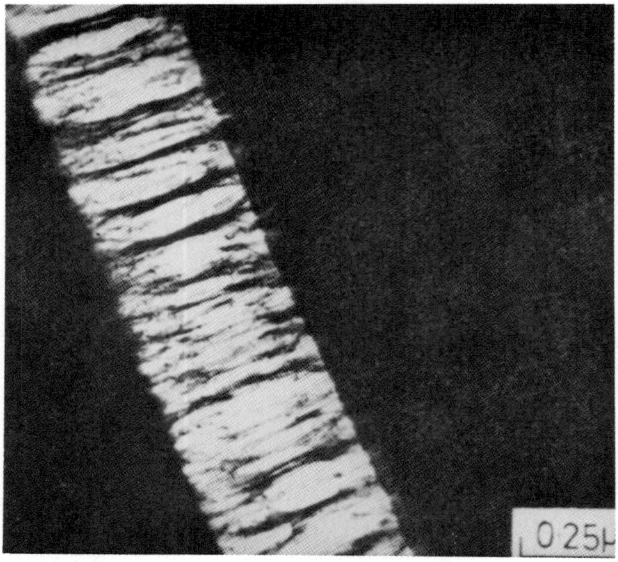

Fig. 98. Craze-band in polystyrene (TEM) (from [86])

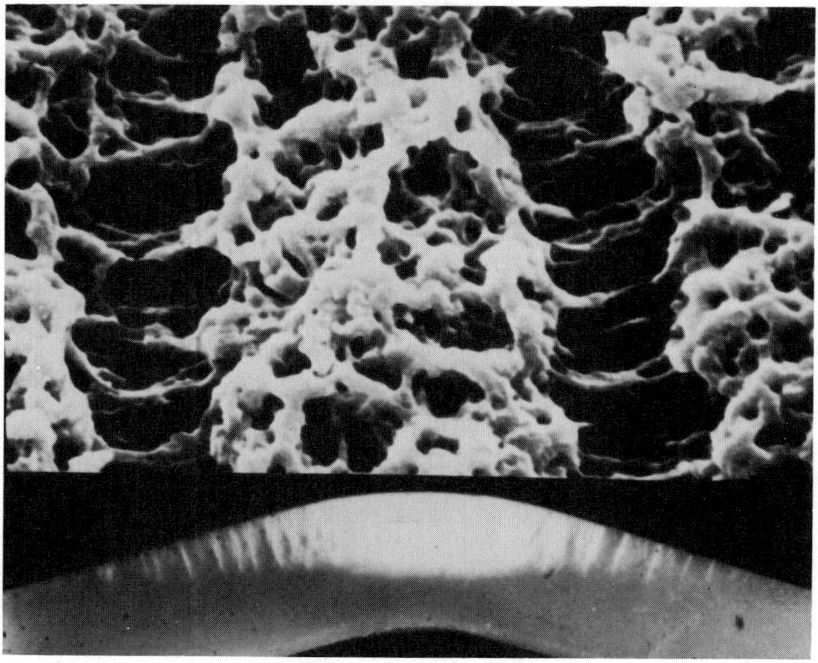

Fig. 99. Crazing with stress-whitening in a sample of ABS after bending (from [87])

deformation is thus, an inhomogeneous deformation involving a stretching and simultaneous opening up of the structures. Associated with this process is a change in the optical properties of the material so that the deformed regions appear white.

This phenomenon is called "stress-whitening" and can be seen particularly well in ABS, a sample of which is shown in Fig. 99. The upper part of Fig. 99 is an electron micrograph of the surface of the sample (after removal of the polybutadiene particles by etching with chromic acid) where bending was most severe. Transoms composed of stretched SAN and orientated normal to the direction of the furrows can be clearly seen.

In order that the elastomer particles can initiate a craze they must be so attached to the matrix that a transfer of force across the phase boundaries is possible. In ABS this is achieved by chemically grafting a shell of the matrix material onto the rubber particles. As a consequence of this coupling and the different thermal expansion coefficients of the matrix and the inclusions, thermal stresses are generated as the material cools from the melt or if it is cooled below ambient temperature. These stresses influence the glass transition of the rubber phase. The phase-coupling resulting from the grafted envelope and the associated T_g-shift are the subject of detailed, fundamental investigations carried out by L. Morbitzer et al. In Fig. 100 the temperature-dependent modulus curves for an ABS graft-copolymerisate and the two individual components (SAN and polybutadiene (PBD)) are reproduced.

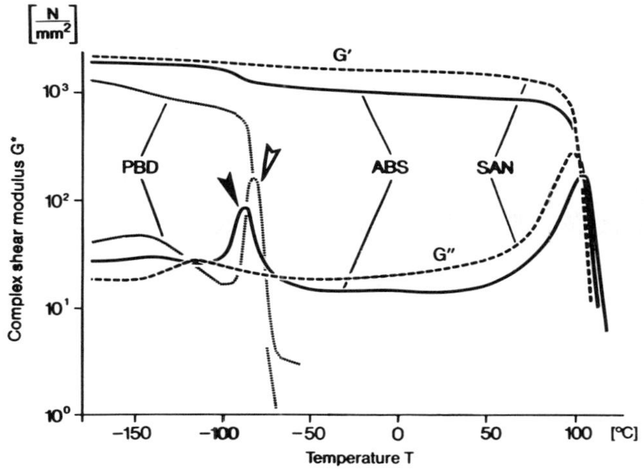

Fig. 100. Temperature dependence of the shear modulus for an ABS graft-copolymerisate and the starting materials (PBD and SAN) (from [85])

During cooling, the thermal stresses are weaker than the interphase bonds and the elastomer particles are hindered from shrinking due to the surrounding, frozen SAN. Thus, the elastomer particles undergo a volume dilation, which induces the T_g of the elastomer phase to shift to temperatures lower than that of ungrafted PBD.

The increase in the T_g of the SAN matrix with respect to that of pure SAN is probably also due to thermal stresses. In the latter case, the elastomer particles impose a pressure on the surrounding matrix which leads to a loss of the segmental mobility and thus raises T_g. With such knowledge it is possible to determine the phase coupling from shifts in the glass transition temperatures for multi-phase systems.

9.3 Shear Deformation

In contrast to craze-deformation, a change of volume is not associated with shear deformation. Characteristic for shear deformation is a pronounced knecking during the uniaxial elongation of a dumbbell test piece. Typical too, are the resulting shear bands orientated diagonally to the direction of the applied stress (Fig. 101).

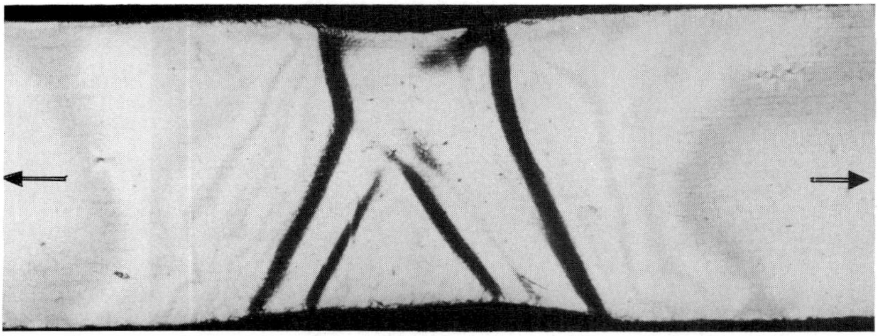

Fig. 101. Shear bands in polycarbonate (from [85])

Craze and shear deformation can, depending on the stress-state, the nature and speed of the applied stress and the temperature, take place simultaneously in a single polymer. Generally speaking, greater amounts of energy are absorbed by materials which undergo shear deformation than by materials which tend to craze.

However, until now the precise structural, polymer specific prerequisites for one mechanism or the other have not been unambiguously defined.

9.4 Deformation Mechanisms in Partially Crystalline Thermoplastics

The lamella is the basic building block of partially crystalline thermoplastics from which the supramolecular structures are built. The most important of such structures is the spherulite (see also Sect. 7.1). In considering the deformation of spherulitic materials it is important to differentiate between homogeneous and inhomogeneous processes. Homogeneous processes are those during which the essentially spherical spherulites are transformed into ellipses. However, after the yield maximum is reached, deformation is predominantly inhomogeneous and micro-cracks often appear along the spherulite interfaces. Since inhomogeneous deformation of thermoplastics can often continue to several hundred percent elongation (see Fig. 93, curve 5) special models have been developed to explain the nature of the mechanisms involved. One possibility for the mechanism of inhomogeneous deformation in partially crystalline thermoplastics has been suggested by J. Schultz and is reproduced in Fig. 102, where two lamellae separated by an amorphous region (*a*) can be seen. Initially, an applied stress leads to an uncoiling and stretching of the chains in the amorphous region (*b*). Lastly, the stress is transferred to the lamellae from the amorphous chains and the former turn to become orientated in the direction of the applied stress (*c*). A further increase in the applied stress leads to individual fold blocks being pulled out of the lamellae and, in the extreme, to the building of separate fibrillar structures (*e*).

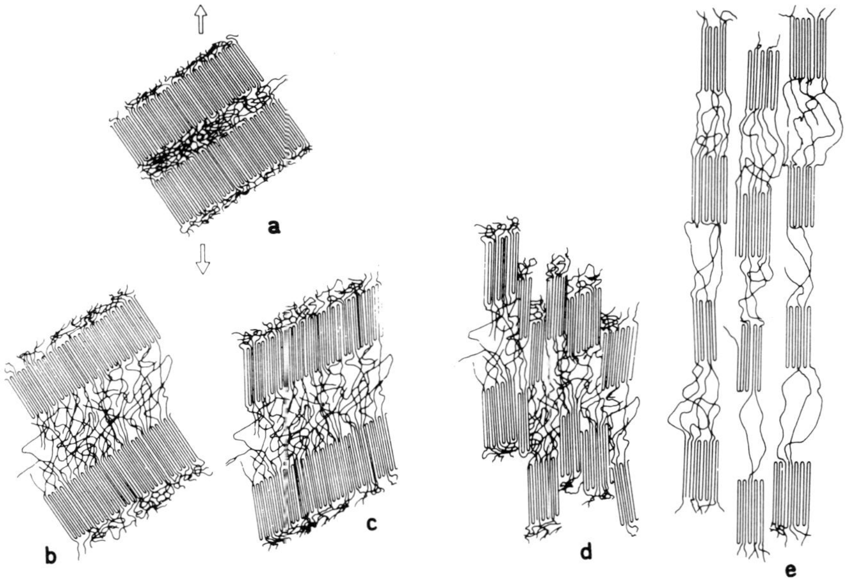

Fig. 102. Model of the stretching mechanism for partially crystalline polymers (from [94])

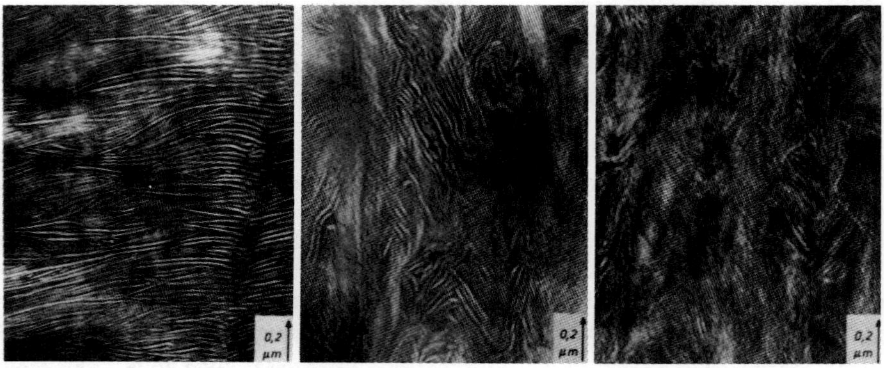

Fig. 103. Electronmicrographs (TEM) of microtome slices from an unstretched PE-fiber (*left*) and from a similar fiber after being stretched to 400 (*middle*) and 600% (*right*) elongation (from [95])

Electronmicrographs of microtome slices from an unstretched PE-fiber and after stretching to 400 and 600% elongation, as shown in Fig. 103, do not lend complete credence to this model.

In the left-hand picture of Fig. 103 the individual lamellae can be seen to be orientated almost normal to the direction of the unstretched fiber. The picture to the right of Fig. 103, after the fiber had been stretched to some six times its original length, cannot, however, be identified in terms of any of the stages of Fig. 102 since the boundaries of the lamellae have become totally blurred.

10 Rubber Elasticity of Covalently Crosslinked Elastomers [98–129]

10.1 Thermodynamics of Rubber Elasticity

Since rubber elastic deformation behavior results, to a first approximation, from reversible change of site processes of chain segments, the first law of thermodynamics can be used as a starting point for a consideration of this phenomenon: The internal energy U of a system can be increased by introducing energy from outside the system, either in the form of thermal energy or as a result of mechanical or electrical work:

$$dU = dQ + dA \tag{183}$$

In order to thermodynamically describe equilibria, variables of state are introduced. An isothermal, isobaric system will tend towards a minimum Gibbs free energy G:

$$G = U - TS + pV \tag{184}$$

$$dG = 0 \text{ with T and p constant at thermal equilibrium} \tag{185}$$

An isothermal, isochoral system tends towards a minimum Helmholtz energy F:

$$F = U - TS \tag{186}$$

$$dF = 0 \text{ with T and V constant at thermal equilibrium} \tag{187}$$

Thus, all processes in an isothermal, isochoral system tend to a minimum internal energy and a maximum entropy.

A single molecule (for example in solution) coils up on itself because the resulting increase in entropy is greater than the simultaneous increase in energy. If a complete system of molecules is considered, it is necessary to take into account the interactions between the individual molecules. For macromolecules with higher orders of symmetry a configuration of minimum potential energy is achieved by cancelling of interaction forces i.e. by crystallization. Molecular asymmetry strongly hinders the ability to crystallize and the material will tend to a maximum entropy configuration. In the latter case the material is amorphous. It is because solid polymers are rarely in thermodynamic equilibrium that both crystalline and amorphous phase can coexist in one and the same sample.

Consider a macromolecular system at constant T and V; the appropriate variable of state is:

$$F = U - TS \tag{188}$$

A change in the Helmholtz energy, brought about by the application of work $dA = f dl$, can be written:

$$dF = dU - TdS = dA = fdl \tag{189}$$

rearrangement gives:

$$dU = dA + TdS = fdl + TdS \tag{190}$$

and from Eq. 190 the force f is given by:

$$f = \left(\frac{\partial U}{\partial l}\right)_{T,V} - T\left(\frac{\partial S}{\partial l}\right)_{T,V} \tag{191}$$

Furthermore, according to Eq. 189:

$$\left.\begin{aligned} \left(\frac{\partial F}{\partial l}\right)_{T,V} &= f \\ \left(\frac{\partial F}{\partial T}\right)_{l,V} &= -S \end{aligned}\right\} \left(\frac{\partial S}{\partial l}\right)_{T,V} = -\left(\frac{\partial f}{\partial T}\right)_{l,V} \tag{192}$$

is valid. Thus, Eq. 191 can also be written as:

$$f = \left(\frac{\partial U}{\partial l}\right)_{T,V} + T\left(\frac{\partial f}{\partial T}\right)_{l,V} \tag{193}$$

As a first approximation, it can be assumed that during a rubber elastic deformation the internal energy of the system remains constant (i.e., no energy elasticity) and that the work done is converted totally, and reversibly, into heat. Equation 193 can now be simplified to yield:

$$f = T\left(\frac{\partial f}{\partial T}\right)_{l,V} = cT \tag{194}$$

The resilience of an elongated rubber sample is thus proportional to the temperature at which the elongation takes place with the consequence that such a sample under constant load will contract if its temperature is increased. In this case pure entropy elasticity is operating.

An experimental check of Eq. 194 requires that the volume of the sample is held constant so that a variable hydrostatic pressure would be necessary. This is very difficult to realize and experiments to observe these effects are generally carried out with variable volume i.e., constant pressure. During such experiments the socalled thermoelastic inversion can be observed at about 10% elongation. Below this inversion the force decreases with increasing temperature; above this inversion force increases with temperature. This anomaly can be ascribed to changes in the volume with temperature: An increase in the reference length of the sample l_0 reduces the elongation $\lambda = l/l_0$ at constant l. Thus, the stress decreases although the modulus indeed increases proportional to the temperature. For small elongations (see also Sect. 10.5):

$$\sigma = 3NkT\varepsilon \tag{195}$$

whereby the elongation can be written in terms of:

$$\varepsilon = \varepsilon_0 - \frac{\beta}{3}(T - T_0) \tag{196}$$

where β is the coefficient of volume expansion. The elongation at the thermoelastic inversion ε_0 can be obtained from $(\partial f/\partial T)_l = 0$ and using Eqs. 195 and 196 one obtains:

$$\varepsilon_0 = \frac{\beta}{3}(2T - T_0) \tag{197}$$

With the characteristic values: $\beta = 6{,}6 \cdot 10^{-4}\,K^{-1}$, $T_0 = 293\,K$ and $T = 343\,K$ the thermoelastic inversion occurs at the same elongation as experimentally observed, i.e, $\varepsilon_0 = 0{,}086$ or 8,6%. At this elongation the temperature coefficient of the stress $(\partial f/\partial T)_l$ changes its sign, agreeing satisfactorily with experiment.

If the energy elastic effects operating at small elongations are explained only in terms of changes in volume, their origin lies in the intermolecular interactions between the polymer chains. On the contrary, Flory, for example, postulated that energy-terms make a contribution to the stress even when the volume remains constant; i.e., the origins lie essentially in intramolecular processes such as exchanges between *trans* and gauche positions in terms of rotational potential.

The energy elastic portion of the force f_e/f can be written in terms of:

$$\left(\frac{\partial U}{\partial l}\right)_{V,T} = f - T\left(\frac{\partial f}{\partial T}\right)_{V,l} \equiv f_e \tag{198}$$

$$\frac{f_e}{f} = -T\left(\frac{\partial \ln (f/T)}{\partial T}\right)_{V,l} \tag{199}$$

In order to adapt the theoretical equations to the experimental conditions (i.e., constant hydrostatic pressure) the Helmholtz energy F can be exchanged for the Gibbs free energy term $G=U+pV-TS$, i.e., $H=U+pV$ is substituted for U. With these substitutions Eq. 198 becomes:

$$\left(\frac{\partial H}{\partial l}\right)_{p,T} = f - T\left(\frac{\partial f}{\partial T}\right)_{p,l} \tag{200}$$

Flory derived the following equation relating the temperature coefficients of stress at constant volume and pressure:

$$\left(\frac{\partial f}{\partial T}\right)_{p,l} = \left(\frac{\partial f}{\partial T}\right)_{V,l} - \frac{f\beta}{\lambda^3 - 1} \tag{201}$$

The quantity λ in this equation corresponds to the strain l/l_0, where l_0 is the length of the unstretched sample with volume V and pressure p. In analogy to Eq. 199, the energy elastic contribution is then given by:

$$\frac{f_e}{f} = -T\left(\frac{\partial \ln (f/T)}{\partial T}\right)_{p,l} - \frac{\beta T}{\lambda^3 - 1} \tag{202}$$

Table 7. Energy Elastic Contribution f_e/f for Several Elastomers (from [98])

Polymer	Reference	f_e/f	Method
Natuaral Rubber	Allen & Coworkers	0.2	const. V
	Allen & Coworkers	0.12 ± 0.02	const. V
	Roe and Krigbaum	0.11 to 0.25 (def. dependent)	const. p
	Ciferri	0.18	const. p
	Shen	0.15 ± 0.3	const. p
	Boyce and Treloar	0.126 ± 0.016	torsion
Butyl rubber	Allen & Coworkers	–0.08	const. V
	Ciferri & Coworkers	–0.03 ± 0.02	const. p
Silicon rubber	Price & Coworkers	0.25 ± 0.01	const. V
Polyethylene	Ciferri	–0.42 ± 0.05	const. p
Polybutadiene	Ref [109]	0.08…0.20	
SBR 15% Styrene	Ref [109]	–0.13	
24% Styrene	Ref [109]	–0.12	
NBR (50% ACN)	Ref [109]	0.03	

For an experiment designed to determine the temperature coefficient of stress at constant strain the following equations can be applied:

$$\left(\frac{\partial f}{\partial T}\right)_{p,\lambda} = \left(\frac{\partial f}{\partial T}\right)_{V,l} + \frac{f\beta}{3} \quad \text{or} \tag{203}$$

$$\frac{f_e}{f} = -T\left(\frac{\partial \ln (f/T)}{\partial T}\right)_{p,\lambda} + \frac{\beta T}{3} \tag{204}$$

The energy elastic contribution f_e/f can be determined experimentally by measuring the temperature coefficient of the stress; Table 7 lists some such values.

10.2 Statistics of the Segment Model

It is convenient to replace the actual molecular structure by an idealized random chain, in which the direction in space of any particular link is entirely random and independent of the orientation of the neighboring links. The links can be envisaged as stiff rods of a length a. Figure 104 shows one possibility of representing such a model chain. The start of the chain is considered to be at the center of the coordinate system.

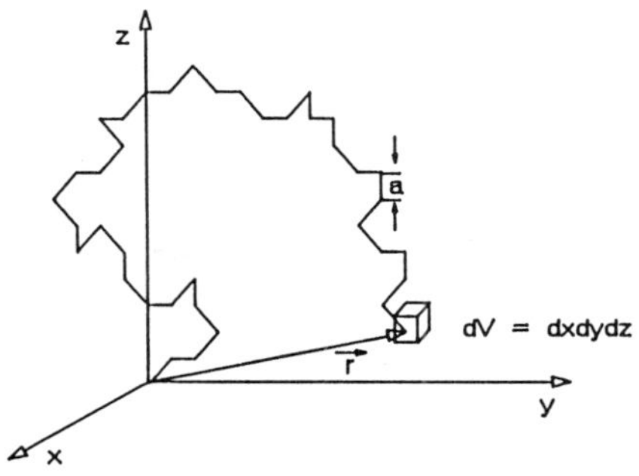

Fig. 104. Random chain model of a polymer molecule

The number of segments per chain is put equal to n and with the number of monomer units per chain segment equal to s, the number of segments is given by:

$$n = P/s \tag{205}$$

where P is the degree of polymerization.

The first problem is to define the most probable length r between the start and the end of the molecular chain given a particular n and a. Generally, a given length r will occur more often the more possible configurations lead to this length. Thus, for example, the maximum length R = na can only be achieved if the chain is fully stretched. Between the extremes (R = na, R = 0) the various lengths R are attained by a much larger number of possible configurations so that these are statistically preferred.

According to Kuhn, the probability w of finding the chain end at the point x, y, z in the unit volume dV = dxdydz is given by (see Fig. 105):

$$w(x, y, z)dV = \left(\frac{b}{\sqrt{\pi}}\right)^3 e^{-b^2(x^2+y^2+z^2)}dV \quad (206)$$

with $b^2 = 3/2na^2$.

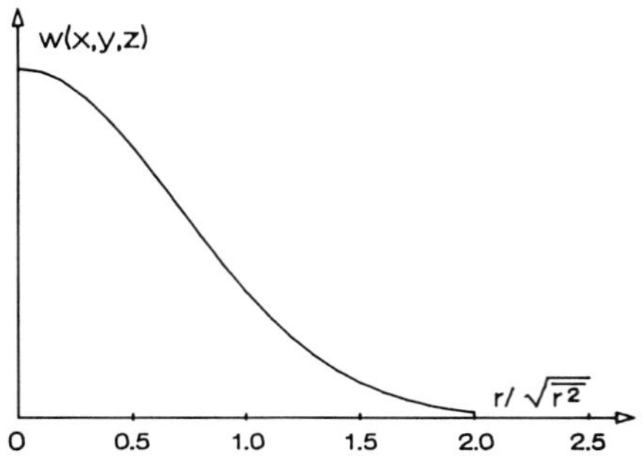

Fig. 105. Density distribution of the chain segments with the chain start as reference point

According to this treatment, the probability of finding the chain end in a volume element of a particular size is greater the closer this element is to the start of the chain. This is not true for a particular length r since the value of r is not only given by the end of the chain being in the volume element dV but, more generally, by the end lying in the spherical envelope between the radii r and r + dr. The probability that this is the case is given by:

$$w(r)dr = \left(\frac{b}{\sqrt{\pi}}\right)^3 e^{-b^2r^2}\, 4\pi r^2 dr = \frac{4b^3}{\sqrt{\pi}}\, e^{-b^2r^2} r^2 dr \quad (207)$$

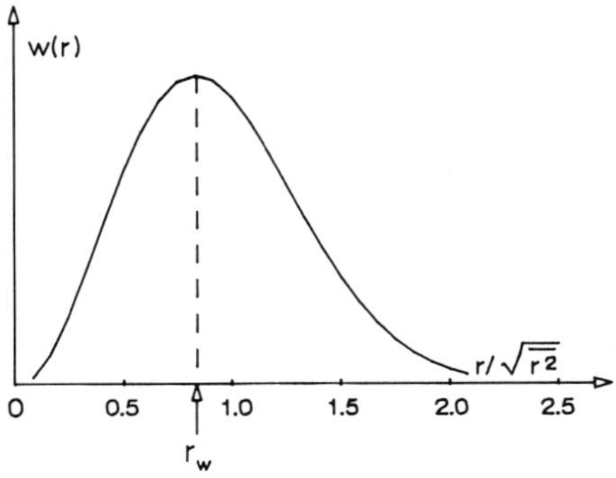

Fig. 106. Plot of the distribution function, Eq. 207, for the end to end distance of a random chain

The most probable length r between the chain ends (the maximum of the curve in Fig. 106) is given by:

$$r_w = 1/b = \sqrt{2n/3}\, a \tag{208}$$

and the mean square of this distance can be derived from:

$$\overline{r^2} = \int_0^\infty r^2 w(r) dr = 3/2b^2 = na^2 \tag{209}$$

so that the average distance between the chain ends is:

$$\sqrt{\overline{r^2}} = \sqrt{n}\, a \tag{210}$$

The average distance between the chain ends increases with the square root of the number of chain segments i.e., with the root of the degree of polymerization P. Furthermore, since the length of the stretched chain R increases linearly with P, the degree of coiling Q for a statistical chain is given by:

$$Q = R/\sqrt{\overline{r^2}} = na/\sqrt{n}\, a \sim \sqrt{P} \sim \sqrt{M} \tag{211}$$

Thus, the statistical filament molecule coils in on itself more and more as its molar mass increases.

10.3 Statistics of Chains with Free Rotation Around their Bond Angles

Real molecules differ from the model of a statistical chain in at least one important respect: The covalent bonds can only rotate within the cone described by the appropriate bond angle. If every orientation on the conical surface is equally probable then the chain is described as a chain with complete freedom of rotation around the bond angles (Fig. 107).

For a paraffin chain the bond angle is 109.5°, the supplement $\alpha = 70.5°$.

If a large enough number of chain elements N with lengths l are considered, then the average of the square of the chain length is given by:

$$\overline{r_1^2} = l^2 N \frac{1 + \cos \alpha}{1 - \cos \alpha} \tag{212}$$

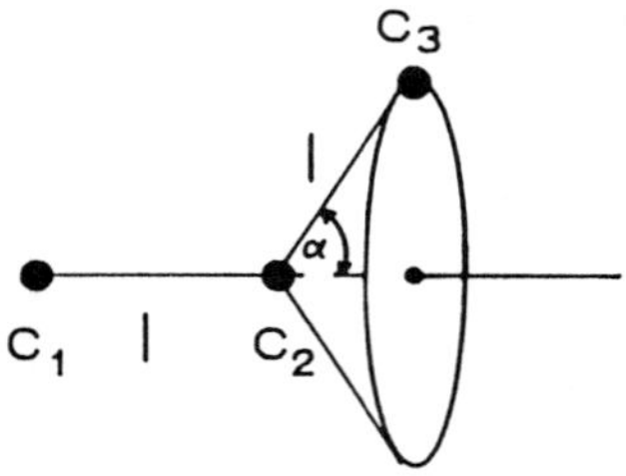

Fig. 107. Rotation of a chain segment on a conical surface

For a C-C-bond (or for other bonds) with a bond angle of 109.5° $\cos\alpha = 1/3$ so that:

$$\overline{r_l^2} = 2l^2N \tag{213}$$

Assuming that the chain length distribution conforms to a Maxwellian rate distribution (Eq. 207) a statistical chain equivalent to real macromolecules can be constructed using the following conditions:

$$\overline{r_l^2} = \overline{r^2} = na^2 \quad \text{and} \tag{214}$$

$$R_l = R = na \tag{215}$$

From these equations the length of a statistical chain segment a and the number of segments n can be calculated provided α and l are known.

Examples (from [98])

a) Chains derived from a simple α-olefin, e.g. paraffin

$\alpha = 70{,}5°;\ l = 0{,}154$ nm

$$\left.\begin{array}{ll} R_l = Nl\cos(\alpha/2) & = na \\ \overline{r_l^2} = 2\,Nl^2 & = na^2 \end{array}\right\} \begin{array}{l} a = 2{,}45\ l \simeq 0{,}38 \text{ nm} \\ n = N/3 \end{array}$$

A statistical chain segment thus composes three C-C-bonds of the actual molecule.

b) Chains derived from a diene, e.g. *cis*-polyisoprene

A single isoprene unit contains 4 covalent bonds: $N = 4n_i$.
According to Wall's numeric evaluation follows:

$$\sqrt{\overline{r^2}} = 2{,}01\sqrt{N} = 4{,}02\sqrt{n_i}$$

with N: Total number of bonds including double bonds
and n_i: Number of isoprene units.

With Eq. (214) and (215) and $l_i = 0{,}44$ nm, the length of an isoprene unit, one obtains:

$$\left.\begin{aligned} \overline{r^2} &= 16{,}2\ n_i = na^2 \\ R_l &= l_i n_i \quad = na \end{aligned}\right\} \quad a = 0{,}368\ \text{nm},\ n \simeq 1{,}2\ n_i$$

Thus, 0.8 isoprene units are equivalent to a statistical chain-segment.

10.4 Statistics of a Covalent Chain with Hindered Rotation Around the Bonds

For C-C chains the rotation around the tetrahedral bond angle is hindered by a rotational potential with a three-fold symmetry. For butane, Volkenstein proposed the following equation for the rotational potential (see also Fig. 108):

$$U(\phi) = \frac{U_0}{2}(1 - \cos 3\phi) + \sum_{i,k} U_{ik}(r_{ik}) \qquad (216)$$

where the subscripts i and k refer to the individual atoms.

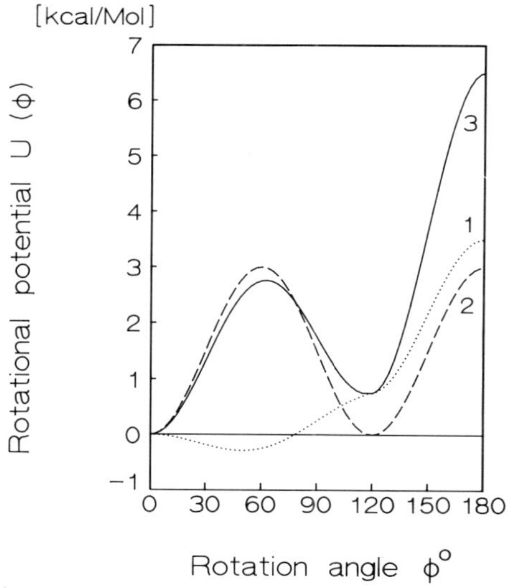

Fig. 108. Rotational potential of butane; *1:* steric potential, *2:* portion due to exchange interactions between neighboring C-H bonds, *3:* total potential

Equation 216 can, in many cases, be simplified to:

$$U(\phi) = \frac{U_0}{2}(1 - \cos 3\phi) + \frac{2}{3}\Delta U(1 - \cos \phi) \tag{217}$$

$$\text{where } U_0 = \begin{cases} 2{,}5 \text{ kcal/mol } (0 < |\phi| < 2\pi/3) \\ 5{,}5 \text{ kcal/mol } (2\pi/3 < |\phi| < \pi) \end{cases}$$

and $\Delta U = 0{,}75$ kcal/mol

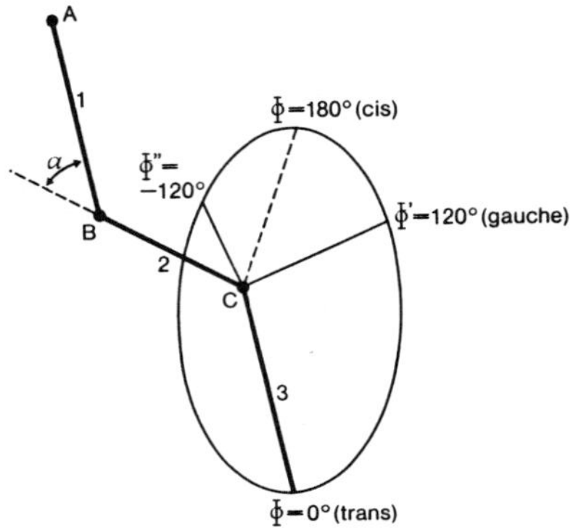

Fig. 109. Establishing the rotation angle about a covalent bond B-C

The probability that the angle ϕ for element 3 ($\phi = 0$ for *trans* configuration, see Fig. 109) is given, according to Boltzmann, by:

$$w(\phi)d\phi = \frac{e^{-U(\phi)/kT}}{\int_{-\pi}^{+\pi} e^{-U(\phi)/kT} d\phi} d\phi \tag{218}$$

From this it follows that the average value of the cosine of the azimuth angle is given by:

$$\eta = \overline{\cos \phi} = \frac{\int_{-\pi}^{+\pi} e^{-U(\phi)/kT} \cos \phi d\phi}{\int_{-\pi}^{+\pi} e^{-U(\phi)/kT} d\phi} \tag{219}$$

Finally, using this average value, the average of the square of the chain length for large enough values of N is given by:

$$\overline{r^2} = l^2 N \left(\frac{1 + \cos\alpha}{1 - \cos\alpha}\right)\left(\frac{1 + \eta}{1 - \eta}\right) \tag{220}$$

Equation 220 can be rewritten in the form:

$$\overline{r^2} = l'^2 N \quad \text{and} \quad l'^2 = l^2 \left(\frac{1 + \cos\alpha}{1 - \cos\alpha}\right)\left(\frac{1 + \eta}{1 - \eta}\right) \tag{221}$$

where l' is the effective length of a bond or a segment of an equivalent statistical chain having N elements which are completely free to rotate.

Since $\overline{\cos\phi}$ is temperature dependent, the end to end distance of a chain with hindered rotation changes with temperature. With increasing temperature, rotation becomes less hindered and the average chain length (with $U \ll kT$) tends towards the limit for the temperature independent freely rotating chain.

10.5 Statistical Theory of Rubber Elasticity

A better understanding of rubber elasticity can be achieved by first answering the following question: How much work dA is required in order to lengthen a statistical chain of length r by an amount dr? Generally, from the first law of thermodynamics (Eq. 183):

$$dA = f dr = dU - T dS, \tag{222}$$

assuming that

$$dU = 0, \tag{223}$$

then

$$f dr = -T dS; \tag{224}$$

and

$$S = k \ln w = C - k b^2 r^2; \tag{225}$$

or

$$dS = -2 k b^2 r dr. \tag{226}$$

In Eq. 225 Boltzmann's definition of entropy in statistical thermodynamic terms is used: The entropy S is proportional to the logarithm of the probability w that a thermodynamic state exists. Boltzmann's constant k is equal to the universal gas constant per molecule. The probability that a particular chain end separation has the value r is given by Eq. 206. Substitution for b in terms of Eq. 208 allows the force f to be defined:

$$f = 3kTr/na^2 \tag{227}$$

In order to hold the two ends of the chain at a particular distance r, a force f is required to cancel the force generated by the desire of the molecules to coil. Vulcanization leads to covalent bonds between the individual macromolecular chains and a three-dimensional network consisting of N statistical chains per cm^3. In order to derive a quantitative relationship between the network structure and its mechanical properties several assumptions are usually made:

a) All chains end at a crosslink i.e., four chains originate at each crosslink.
b) Cyclization and entanglements are negligible.
c) The chains have zero volume and do not interact.
d) A macroscopic deformation leads to a deformation of the network chains in the same ratio as the corresponding dimensions of the bulk rubber (this is called the affinity condition).

The application of an external, macroscopic stress leads to each chain in the network being forced into a statistically less likely conformation. The tendency of the chains to return to a more probable conformation (maximum entropy) is manifested as a tendency of the macroscopic sample to change its form. This is the resilience which is termed rubber or entropy elasticity.

In order to calculate the elastic constants, a three-dimensional network containing N statistical chains per cm^3 all ending at a crosslink is considered. All the chains are treated as being composed of the same number of segments n so that they are all characterized by the same values for b^2 and r^2 (the assumption of a distribution for n leads to an identical result). Fluctuations of the junctions are not considered. Taking a cube with an edge length 1 and applying a force F in the direction x (see Fig. 110). Furthermore, assume, as a first approximation, that the volume of the cube under such an uniaxial deformation remains constant (incompressability; i.e. $\mu = 0.5$). It follows that:

$$(1 + \varepsilon)(1 + \varepsilon_y)(1 + \varepsilon_z) = 1 \tag{228}$$

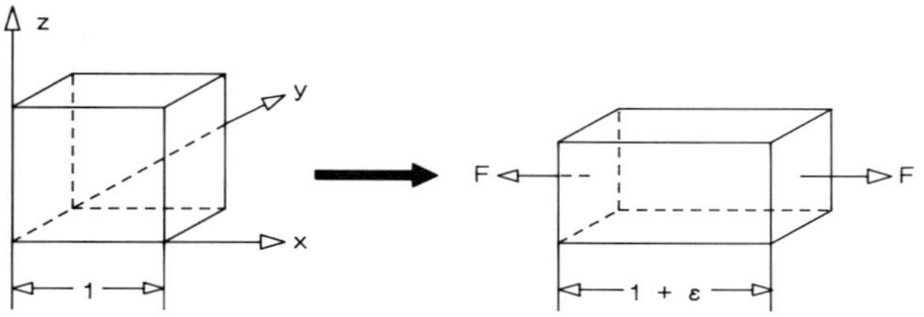

Fig. 110. Stretching of an unit cube

For reasons of symmetry $\varepsilon_y = \varepsilon_z$ and thus:

$$(1 + \varepsilon_y) = (1 + \varepsilon_z) = (1 + \varepsilon)^{-1/2} \tag{229}$$

The distance between two crosslinks $r_i(x_i, y_i, z_i)$ after elongation becomes $r_i'(x_i', y_i', z_i')$ with

$$x_i' = (1 + \varepsilon)x_i \tag{230a}$$

$$y_i' = (1 + \varepsilon)^{-1/2}y_i \tag{230b}$$

$$z_i' = (1 + \varepsilon)^{-1/2}z_i \tag{230c}$$

An expression for the entropy can be obtained from Eqs. 225 and 206 which, after summation for all the chains, leads to:

$$S = C - kb^2 \sum_i r_i^2 = C - kb^2 \sum_i (x_i^2 + y_i^2 + z_i^2) \tag{231}$$

The elongation leads to a reduction in the entropy ΔS given by:

$$S - \Delta S = C - kb^2 \sum_i [(1 + \varepsilon)^2 x_i^2 + (1 + \varepsilon)^{-1} y_i^2 + (1 + \varepsilon)^{-1} z_i^2] \tag{232}$$

Since chain lengths are equally distributed over all three coordinate directions:

$$\sum_i x_i^2 = \sum_i y_i^2 = \sum_i z_i^2 = \frac{1}{3} \sum_i r_i^2 = \frac{N}{3} \overline{r^2} = \frac{N}{2b^2} \tag{233}$$

and thus:

$$\Delta S = -\frac{1}{2} kN \left[(1 + \varepsilon)^2 + \frac{2}{1 + \varepsilon} - 3\right] \tag{234}$$

The amount of work done on the cube is given by:

$$\Delta A = f\Delta r = \Delta U - T\Delta S \tag{235}$$

With the assumption that

$$\Delta U = 0 \tag{236}$$

it follows that:

$$\Delta A = -T\Delta S = \frac{1}{2} kTN \left[(1 + \varepsilon)^2 + \frac{2}{1 + \varepsilon} - 3\right] \tag{237}$$

and with Eq. 237 the force f can be written:

$$f = \frac{\partial(\Delta A)}{\partial \varepsilon} = kTN\left[1 + \varepsilon - \frac{1}{(1+\varepsilon)^2}\right] \tag{238}$$

which leads to the following expression for the stress based on the actual cross-section of the sample:

$$\begin{aligned} \sigma &= \frac{f}{(1+\varepsilon)^{-1}} = kTN\left[(1+\varepsilon)^2 - \frac{1}{1+\varepsilon}\right] \\ &= 3kTN\left[\varepsilon + \frac{1}{3}\frac{\varepsilon^3}{1+\varepsilon}\right] \end{aligned} \tag{239}$$

Considering only elongations for which terms having exponents greater than one can be ignored allows Eq. 239 to be simplified to:

$$\sigma = 3kTN\varepsilon = E\varepsilon \tag{240}$$

where

$$E = 3kTN \tag{241}$$

or, since incompressibility has been assumed:

$$G = E/3 = NkT \tag{242}$$

With $\lambda = l/l_0 = \varepsilon + 1$ dependence of stress on strain (from Eq. 239) is given by:

$$\sigma = NkT(\lambda - \lambda^{-2}) \tag{243}$$

Here σ is based on the cross-section of the unstretched sample. If n is the number of moles of network chains per unit volume:

$$\sigma = nRT(\lambda - \lambda^{-2}) \tag{244}$$

or alternatively, if M_c is the number average molar mass of the network chains ($nM_c = \varrho$):

$$\sigma = (\varrho RT/M_c)(\lambda - \lambda^{-2}) \tag{245}$$

There are two other equations which are often used for the dependence of force on an uniaxial elongation:

$$f = (\nu kT/l_i)(\overline{r_i^2}/\overline{r_0^2})(\lambda - \lambda^{-2}) \quad \text{or} \tag{246}$$

$$f = (\nu kT/l_i)(V/V_0)^{2/3}(\lambda - \lambda^{-2}) \tag{247}$$

Here ν is the number of chains in the network and l_i is the length of the unstretched sample. The term $\overline{r_i^2}/\overline{r_0^2}$ (or $(V/V_o)^{2/3}$) is called the memory term or dilation factor. The term $\overline{r_i^2}$ is the square of the average end to end distance of the network chains in the undeformed network and $\overline{r_0^2}$ is the same quantity for the uncrosslinked chains in an identical environment. Since $\overline{r_i^2} \approx \overline{r_0^2}$, $\overline{r_i^2}/\overline{r_0^2}$ can be put equal to 1.

From Eqs. 242 and 245 the average molar mass M_c of the network chains can be calculated from the modulus of elasticity or the shear modulus. Typical values for a sample vulcanized with a sulfur system are, for example: $E = 1\ N/mm^2$ and $\varrho = 1\ g/cm^3$ which lead to a value of the molar mass at $T = 298$ K of $M_c = 3\varrho RT/E = 7.4 \cdot 10^3$ g/mol.

Flory, in his calculations of network density, includes the number of free chain-ends. Such moieties reduce the elastically effective number of network chains. Flory assumed that a portion of the junctions are required to convert the molecules from their initial, finite length into a single chain of infinite length. The average molar mass before crosslinking is given by M_n and there are ϱ/M_n mol polymer per unit volume. In order to connect all the polymer chains to form one "infinitely long" molecule ϱ/M_n mol crosslinks per unit volume are required. If c is the total number of tetrafunctional crosslinks per unit volume then the number of effective crosslinks per cm^3 is given by:

$$c_e = c - \varrho/M_n \tag{248}$$

Four chains originate at each crosslink and each chain is connected to two crosslinks so that for every effective junction there are two effective network chains. Thus,

$$n_e = 2(c - \varrho/M_n) \tag{249}$$

and since $c = \varrho/2M_c$

$$n_e = \varrho/M_c - 2\varrho/M_n \tag{250}$$

With Eq. 250 the stress-strain relationship of Eq. 244 can now be rewritten as:

$$\sigma = \frac{\varrho RT}{M_c}\left(1 - \frac{2M_c}{M_n}\right)(\lambda - \lambda^{-2}) \tag{251}$$

10.6 Stress-Strain Relationships for Different Types of Applied Stress

10.6.1 Uniaxial Tension or Compression

An elongation λ in a uniaxial direction causes, normal to this direction, the strain components λ_2 and λ_3

$$\lambda_1 = \lambda;\ \lambda_2 = \lambda_3 = \lambda^{-1/2}$$

The resulting equation relating the stress and the strain can be derived directly from Eq. 239:

$$\sigma = G(\lambda - \lambda^{-2}) \quad \text{with} \tag{252}$$

$$G = NkT.$$

For compression the same equation is valid but with $\lambda < 1$.

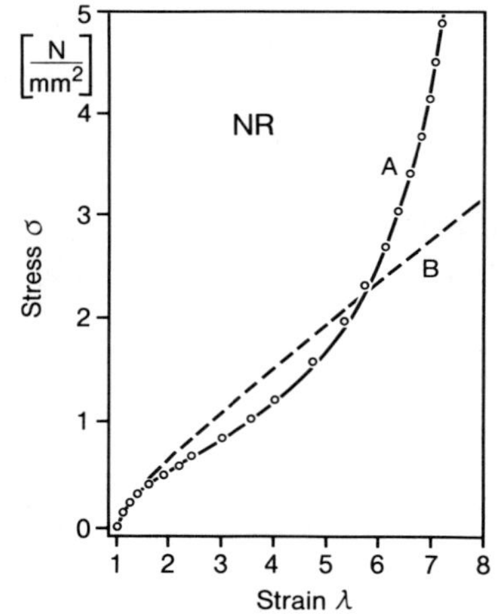

Fig. 111. Uniaxial stress-strain curves for a natural rubber gum-stock
A: Experimentally determined, **B:** Theoretical curve from Eq. 252 with $G = 0.4\ MN/m^2$ (from [98])

Figure 111 compares an experimental, uniaxial stress-strain curve with that calculated from Eq. 252.

10.6.2 Biaxial Elongation

In this case the three main strains are related by:

$$\lambda_1 = \lambda_2 = \lambda;\ \lambda_3 = \lambda^{-2}$$

Thus, the stress-strain relationship results:

$$\sigma = G(\lambda^2 - \lambda^{-4}) \quad (253)$$

10.6.3 Simple Shear

Since during simple shear of a cube, such as that shown in Fig. 112, there is no deformation normal to the *xz*-plane (in the *y*-direction):

$$\lambda_1 = \lambda;\ \lambda_2 = 1;\ \lambda_3 = 1/\lambda \quad \text{and}$$

$$\gamma = \tan\phi = \lambda - 1/\lambda \quad (254)$$

is valid. Generally, the decrease in entropy due to such a deformation of a unit cube is given by:

$$\Delta S = -\frac{1}{2} Nk(\lambda_1^2 + \lambda_2^2 + \lambda_3^2 - 3) \quad (255)$$

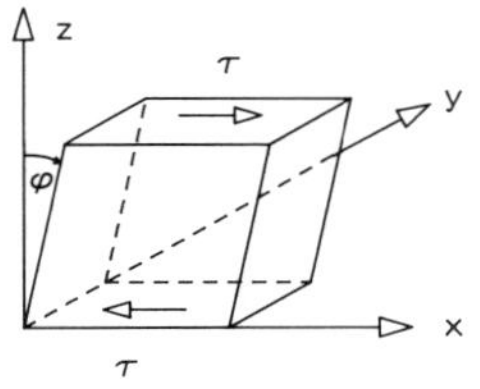

Fig. 112. Simple shear of a unit cube

Putting $\Delta A = -T\Delta S$ and Eq. 254 one obtains, for simple shear, the relationship:

$$\Delta A = \frac{1}{2} NkT(\lambda^2 + \lambda^{-2} - 2) = \frac{1}{2} G\gamma^2 \quad (256)$$

and

$$\tau = \frac{d(\Delta A)}{d\gamma} = G\gamma \quad (257)$$

Thus, for simple shear, in contrast to uniaxial elongation, Hooke's law is valid.

10.7 Phantom Networks

The theory of phantom networks, developed by James and Guth, is based on the same premise as is the affine theory of networks described above; namely, that the elastic free energy is stored by Gaussian network chains (intramolecularly). Intermolecular interactions are assumed to make no contribution to the free energy and the molecules are assumed to be devoid of material properties i.e., they can intertwine freely and do not exclude one another from a volume element. The crosslinks are, however, allowed to fluctuate around their mean positions due to Brownian motion and only these mean positions are shifted in an affine manner by a macroscopic deformation. In contrast, the instantaneous distribution of the chain vectors is not affine with the macroscopic deformation because it is determined not only by the distribution of the mean chain vectors $\bar{r}$ (which are affine) but also by the distribution of the strain-independent fluctuations of the junctions Δr. The stress-strain relationship can be written:

$$\sigma = \frac{\xi kT}{l_i}\left[\frac{V}{V_0}\right]^{2/3}(\lambda - \lambda^{-2}) \tag{258}$$

The parameter ξ here is a characteristic of the network; generally $\xi < \nu$. For a perfect phantom network

$$\xi = \nu(\phi - 2)/\phi \tag{259}$$

holds true. In Eq. 259 ν represents the number of chains in the network and ϕ represents the functionality of the crosslinks. For a network with tetrafunctional crosslinks:

$$\xi = \nu/2 \tag{260}$$

which implies that the retractive force of a phantom network having tetrafunctional crosslinks is only one-half of that of a purely affine network.

A great achievement is the molecular theory developed by Flory. In this model the limitations on the fluctuations of the junctions are represented by domains of constraints. These are formed by the physical entanglements of neighboring chains. Such entanglements should, according to Flory, not be considered as elastically effective crosslinks. The topological and mathematical treatment of a network with domains of constraints leads to the following expression for the modulus, represented here in the form of reduced stress, i.e., based on the appropriate characteristic strain function:

$$[\sigma] = \sigma_{ph}\left[1 + \frac{\sigma_c}{\sigma_{ph}}\right] \tag{261}$$

The term σ_{ph} is that part of the stress corresponding to the phantom network and σ_c represents that part of the stress resulting from the entanglement constraints. The ratio σ_c/σ_{ph} can be written as a rather complicated function which, for the case of uniaxial elongation, has the form:

$$\frac{\sigma_c}{\sigma_{ph}} = \frac{\mu[\lambda K(\lambda^2) - \lambda^{-2}K(\lambda^{-1})]}{\xi(\lambda - \lambda^{-2})} \tag{262}$$

Here, μ is the number of effective crosslinks and ξ is the same network parameter as in Eq. 259. In Flory's terms ξ represents the cycle rank of the network. This is defined as the number of chains which must be cut in order to reduce the network to an acyclic structure or tree.

$$\xi = \nu - \mu \tag{263}$$

The function K in Eq. 262 depends on two additional parameters which take account of the severity of the entanglement constraints relative to those imposed by the phantom network and the non-affine transformation of these domains with strain. Flory's network theory can be seen to form a connection between the affine and phantom network theories. That is, for small elongations the physical entanglements of the macromolecules lead to considerable restrictions of the fluctuations of junctions so that these tend to move in an affine manner. At large strains, or at high dilation, the restrictions to junction fluctuations vanish and the stress to strain relationship approaches that for a phantom network.

10.8 Mooney-Rivlin Theory

The storable, elastic energy of an unstressed, isotropic and incompressible material can, in its most general form, be represented as a function of the strain invariants I_n. These invariants are defined by:

$$I_1 = \lambda_1^2 + \lambda_2^2 + \lambda_3^2 \tag{264a}$$

$$I_2 = \lambda_1^2\lambda_2^2 + \lambda_2^2\lambda_3^2 + \lambda_1^2\lambda_3^2 \tag{264b}$$

$$I_3 = \lambda_1^2\lambda_2^2\lambda_3^2 \tag{264c}$$

As a result of the incompressibility condition ($V = \text{const.}$) $I_3 = 1$ so that the storable elastic energy of the network is only a function of I_1 and I_2: $W(I_1(\lambda_i^2), I_2(\lambda_i^2\lambda_j^2))$. It can be represented by an expansion of the following series:

$$W = \sum_{i,j=0}^{\infty} C_{ij}(I_1 - 3)^i (I_2 - 3)^j \tag{265}$$

For $\lambda = 1$, $I_1 = I_2 = 3$ and $W = 0$ so that C_{00} is also zero.

A first approximation is thus given by:

$$\begin{aligned} W &= C_{10}(I_1 - 3) + C_{01}(I_2 - 3) \\ &= C_{10}(\lambda_1^2 + \lambda_2^2 + \lambda_3^2 - 3) + C_{01}(1/\lambda_1^2 + 1/\lambda_2^2 + 1/\lambda_3^2 - 3) \end{aligned} \tag{266}$$

For an uniaxial elongation:

$$\lambda_1 = \lambda;\ \lambda_2 = \lambda_3 = \lambda^{-1/2}$$

which leads to ($C_{10} \equiv C_1$, $C_{01} \equiv C_2$)

$$W = C_1(\lambda^2 + 2/\lambda - 3) + C_2(\lambda^{-2} + 2\lambda - 3) \tag{267}$$

and

$$\sigma = \frac{dW}{d\lambda} = C_1(2\lambda - 2\lambda^{-2}) + C_2(2 - 2\lambda^{-3}) \tag{268}$$

or

$$\sigma = 2(C_1 + C_2/\lambda)(\lambda - \lambda^{-2}) \tag{269}$$

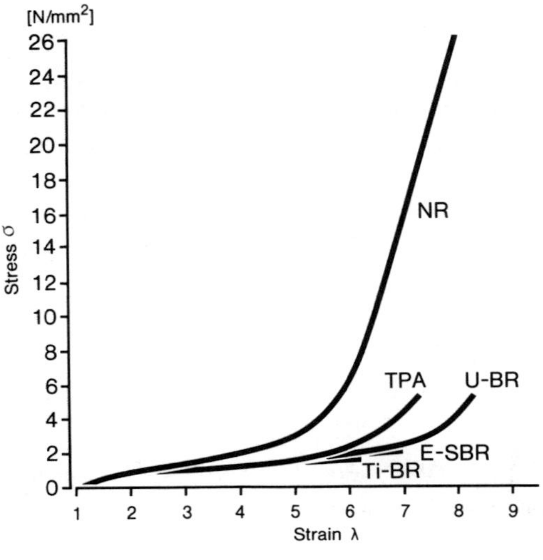

Fig. 113. Stress-strain curves of some elastomers

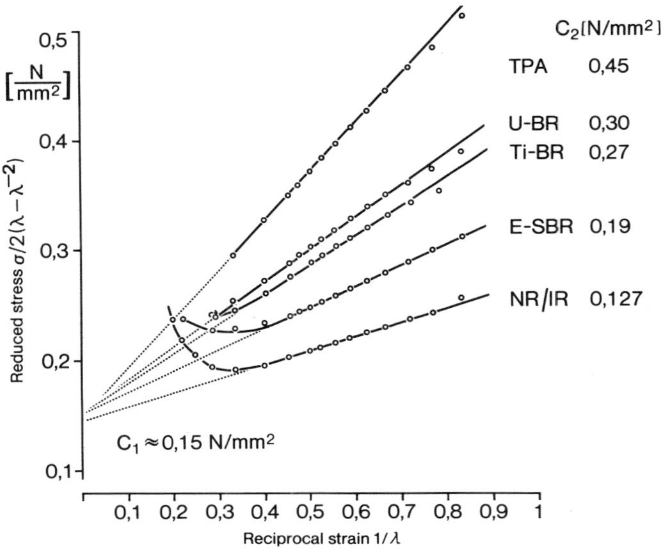

Fig. 114. Mooney-Rivlin plots of the stress-strain curves shown in Fig. 113 according to Eq. 269

The latter is the well known Mooney-Rivlin equation which is equivalent to the simple equation of a Gaussian network if $C_2 = 0$. In this case (see Eq. 252)

$$2C_1 = G \qquad (270)$$

A plot of the reduced stress $\sigma/2(\lambda-\lambda^{-2})$ as a function of reciprocal elongation $1/\lambda$ gives a straight line whose slope is C_2 and whose intercept with the ordinate is C_1. In Fig. 113 stress-strain curves for several elastomers are shown which are converted into Mooney-Rivlin plots (Eq. 269) in Fig. 114.

In practice the constant C_1 has proved to be a useful measure of the crosslink density. Thus, for both sulfur and radiation crosslinked elastomers there is a linear

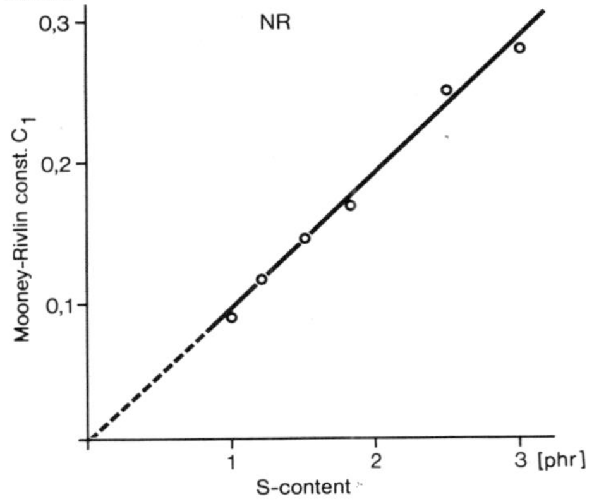

Fig. 115. Dependence of the Mooney-Rivlin constant C_1 on the concentration of sulfur

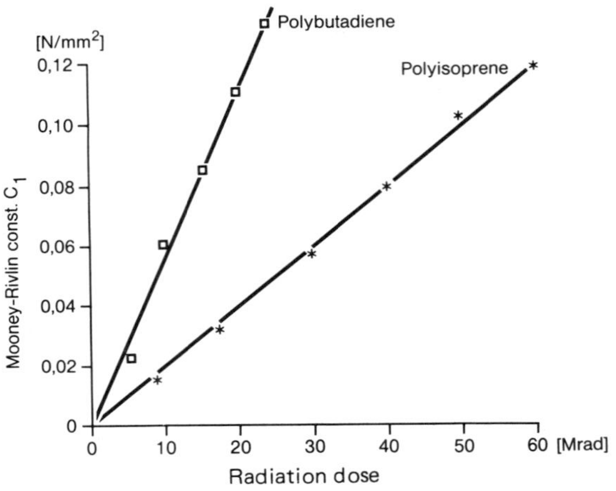

Fig. 116. Dependence of the Mooney-Rivlin constant C_1 on the radiation dose

dependence of C_1 on the concentration of sulfur or the radiation dosage (see Fig. 115 and 116).

In many cases, an extrapolation to $C_1 = 0$ is not allowed since in the absence of chemical crosslinks the physical entanglements also contribute to C_1.

The crosslink density of filled systems can also, as suggested by W. L. Hergenrother, be analyzed with the aid of certain assumptions using the Mooney-Rivlin equation.

In order that Eq. 269 can be applied to filled elastomers, these must exhibit entropy elastic deformation behavior. However, especially highly active fillers lead to deviations from merely entropic elasticity. The main reasons for this are the interactions between the filler particles and the polymer chains and between the individual filler particles both of which are superimposed on the covalent network.

The individual filler particles, especially of highly active carbon blacks, initiate secondary and tertiary structures which lead to an overlying network. It is, however, possible to mechanically destroy this network formation of carbon black without destroying the chemical network. The breaking down of the "filler network" and the slip processes at the filler/polymer interfaces can be observed during cyclical stress/strain experiments as a stress softening of the sample (also called the Mullin's effect, see Fig. 117). In the case of filled rubber stocks, the macroscopic strain λ in the Mooney-Rivlin equation must be replaced by an intrinsic strain. This is necessary, because the cross-section of the highly elastic network is reduced due to the proportion of the hard, inelastic filler. Thus, if the volume of the total system remains constant, the strain of the individual network chains must increase. The corresponding strain amplification factor

$$\Lambda = \varepsilon x + 1 \tag{271}$$

is calculated using an empirical factor x which, in turn, can be obtained, e.g. from modulus measurements of both filled and unfilled systems. A filled system composed of a dispersion of spherical particles was first mathematically described by Einstein who recognized that the viscosity increases linearly with an increasing concentration of dispersed particles. Einstein's initial equation has been expanded to include a quadratic term by Guth and Gold and has been used to rationalize the moduli of filled polymer systems by, among others, Smallwood and Mullins. From these theories the factor x in Eq. 271 is given by:

$$x = 1 + 2{,}5\, v_f + 14{,}1\, v_f^2 \tag{272}$$

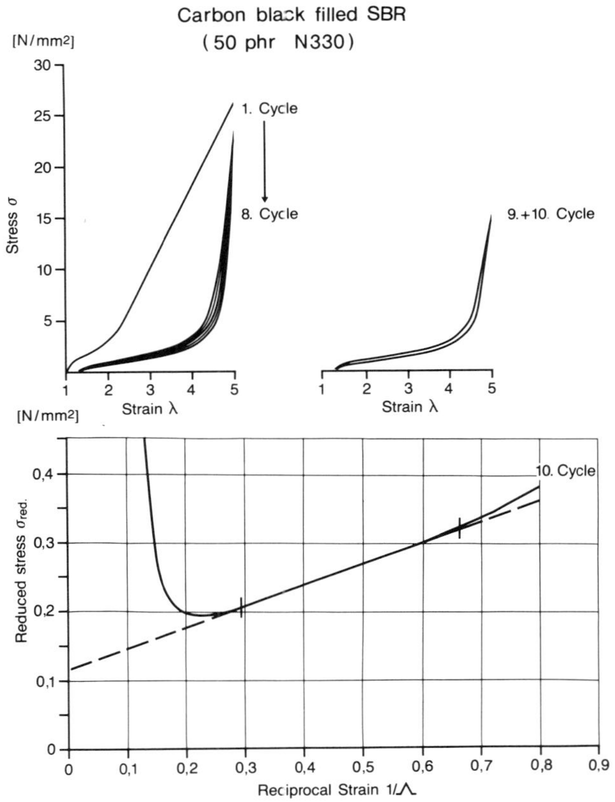

Fig. 117. Stress softening of a SBR-elastomer filled with carbon black after cyclic deformation (*top*) and the Mooney-Rivlin plot after ten cycles (*bottom*)

Figure 117 shows an example of a Mooney-Rivlin evaluation (*bottom* picture) of a stress-strain curve obtained from a filled elastomer after the breakdown of the carbon black "network" by nine cyclic deformations (*top* picture). The stress softening is complete after some nine cycles since the tenth and subsequent stress-strain curves are superimposed on each other. By such a procedure, the C_1 value obtained with the aid of Eqs. 269, 271 and 272 is only slightly different from the crosslink density of an equivalent unfilled network.

Considerably more difficult is the interpretation of the C_2 constant in terms of polymer or network specific characteristics. It has been shown that C_2 decreases

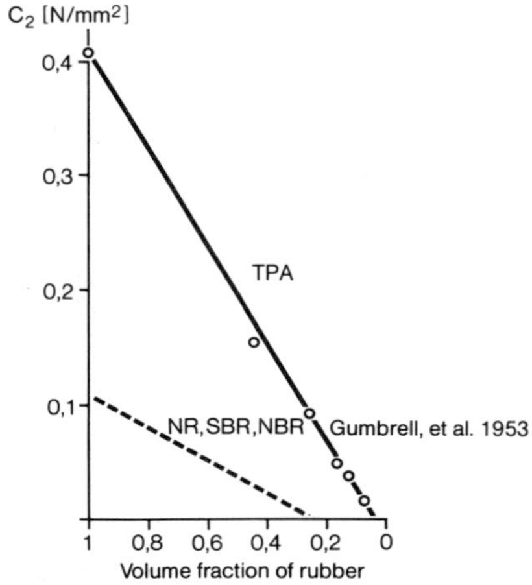

Fig. 118. Dependence of the Mooney-Rivlin constant C_2 on the degree of swelling for several elastomers

with increasing degree of swelling. This effect is shown in Fig. 118. Furthermore, C_2 decreases as the chain cross-section increases, i.e., as the polymer chain becomes stiffer and the conformational entropy of the network chains decreases. Such experiments are used to corroborate the idea that C_2 reflects the concentration of physical, more unstable crosslinks (such as entanglements, filler/filler and filler/polymer interactions). This interpretation is supported by experiments whereby such structures are broken down by swelling or cyclic deformation.

10.9 Non-Gaussian Chain Statistics and Network Theory

As can be seen in Fig. 114, deviations from the straight line predicted by the Mooney-Rivlin theory occur at large elongations (small values of $1/\lambda$). These deviations can be ascribed to the neglect of the anisotropic distribution of the chain vectors, which becomes important at larger elongations due to an orientation of the network. This deficiency is a consequence of the Gaussian chain statistics on which the theory is based.

A starting point for a phenomenological, statistical treatment of larger elongations is the ideal chain with n links of length l. A distribution function is then developed for the angle of the individual links of the chain with respect to a chosen direction with the condition that the chain ends are fixed in this chosen direction. The distribution function w(r) for the chain vectors r can be found provided that w(r) is proportional to the most probable distribution of the segmental angles for the corresponding value of r:

$$\ln w(r) = C - n\left(\frac{r}{nl}\beta + \ln\frac{\beta}{\sinh\beta}\right) \tag{273}$$

In this equation $\beta = L^{-1}(r/nl)$ is the inverse Langevin function. Equation 273 can be expanded as a series to give (from [98])

$$\ln w(r) = C - n\left[\frac{3}{2}\left(\frac{r}{nl}\right)^2 + \frac{9}{20}\left(\frac{r}{nl}\right)^4 + \frac{99}{350}\left(\frac{r}{nl}\right)^6 + \ldots\right] \tag{274}$$

in which the first term (Eq. 275) corresponds to a Gaussian distribution.

$$\ln w(r) = C - \frac{3r^2}{2nl^2} \tag{275}$$

The probability that a chain vector, independent of the direction, lies between r and r + dr can be obtained in the same way as it was in the Gaussian case (Sect. 10.2). That is, by multiplying the probability density from Eq. 273 with the volume element of the spherical envelope $4\pi r^2 dr$.

The entropy of an individual chain can be obtained from:

$$s = k \ln w = C' - kn\left(\frac{r}{nl}\beta + \ln\frac{\beta}{\sinh\beta}\right) \tag{276}$$

The calculation of the entropy for the total network can, in the simplest approach, be carried out using the socalled 3-chain model. This model is based on the assumption that the properties of a Gaussian network can be reduced to those of three independent sets of chains with r-vectors parallel to the coordinate axes and that this approach can also be applied in the non-Gaussian region. If the deformation consists of a simple extension λ in the x-direction the r-vectors of the three chains in the deformed state, assuming incompressibility, will be:

$$r_x = \lambda r_0; \quad r_y = r_z = \lambda^{-1/2} r_0 \tag{277}$$

If s_x, s_y and s_z represent the corresponding entropies of the chains and since $s_y = s_z$ the total entropy elastic stress is given by:

$$\sigma = -T\frac{ds}{d\lambda} = \frac{NT}{3}\frac{d}{d\lambda}(s_x + 2s_y) \tag{278}$$

or

$$\sigma = \frac{NkT}{3}\frac{r_0}{1}\left[L^{-1}\left(\frac{r_0\lambda}{nl}\right) - \lambda^{-3/2} L^{-1}\left(\frac{r_0\lambda^{-1/2}}{nl}\right)\right] \tag{279}$$

Putting $r_0 = ln^{1/2}$ for the free chain produces:

$$\sigma = \frac{NkT}{3} n^{1/2}\left[L^{-1}\left(\frac{\lambda}{n^{1/2}}\right) - \lambda^{-3/2} L^{-1}\left(\frac{1}{\lambda^{1/2} n^{1/2}}\right)\right] \tag{280}$$

The following series expansion can be substituted for $L^{-1}(x)$ (from [98])

$$L^{-1}(x) = 3x + \frac{9}{5}x^3 + \frac{297}{175}x^5 + \frac{1539}{785}x^7 + \ldots \tag{281}$$

from which it can be seen that for small values of x, $L^{-1}(x) = 3x$ so that Eq. 279 becomes:

$$\sigma = NkT(\lambda - \lambda^{-2}) \tag{282}$$

Thus, the chain vectors have a Gaussian distribution. The non-Gaussian approximation for a stress-strain curve according to Eq. 280 is governed by two parameters:

a) N = the number of network chains per unit volume and
b) n = the number of random links in the network chain.

These two parameters can only be treated as mathematically independent. Physically, for a given polymer, they are dependent from each other since N determines the molar mass M_c of the network chains with the result that N is inversely proportional to n. However, since for a given M_c, n is usually unknown, it is not possible to derive a numerical relationship between n and N. Normally, these parameters are obtained empirically by fitting the curve from Eq. 280 to an experimentally obtained stress-strain curve. Examples of this procedure are shown in Fig. 119.

It can be seen that, with the chosen values of $G = NkT = 0.25$ N/mm^2 and $n = 102$, a good fit between the theoretical and experimental curves for SBR can be

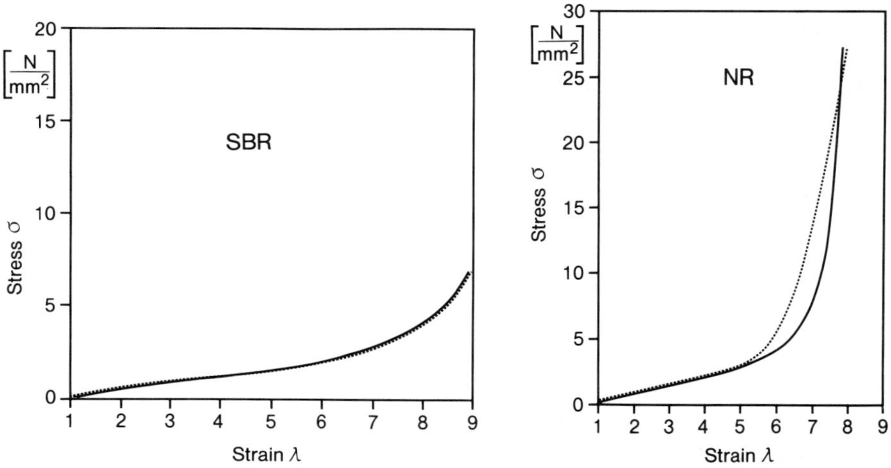

Fig. 119. Comparison of stress-strain curves from unfilled SBR and NR. The *full lines* are experimentally obtained curves; the *dotted lines* are calculated from Eq. 280 with:
$G = 0.25$ N/mm^2 and $n = 102$ for SBR ($T = 0°C$);
$G = 0.4$ N/mm^2 and $n = 67$ for NR ($T = 23°C$)

obtained. In contrast, a similar treatment for a NR gum stock leads to a considerable discrepancy between the theoretical and experimental curves, especially at larger elongations, when the fitting ($G = 0.4\ N/mm^2$ and $n = 67$) is carried out at low strain. The reason for the discrepancy is the strain-induced crystallization of the NR sample, which is not accounted for by the theory and leads to considerably more self-reinforcement than would be caused by the limited extensibility of the network. A description of the deformation behavior of acrylonitrile-butadiene copolymerisates with the aid of the inverse Langevin Equation (Eq. 280) has also been made by M. Hoffmann with considerable success.

10.10 Van der Waals Theory of Networks [124–129]

A novel attempt to describe rubber elastic deformations has been made in terms of the van der Waals equation of state by H.-G. Kilian. According to this approach, an isotropic Gaussian network behaves like an ideal, conformational gas. However, in order to describe the global properties of real networks the volumes, finite lengths and the mutual interactions of the macromolecular chains have to be taken into account. This is done formally, in a manner analogous to that used in formulating the van der Waals equation of state for real gases.

Thus

$$p = \frac{\nu kT}{V - b^*} - \frac{a^*}{V^2} \tag{283}$$

and in analogy:

$$f = \frac{\nu kT}{l_0}\left[\frac{1}{(D^{-1} - b)} - a_0 D^2\right] \tag{284}$$

where ν is the number of gas molecules or network chains in the considered volume V, b^* is the appropriate volume of the gas molecules and a^* takes care of their interactions. In addition, D corresponds to the characteristic strain function $(\lambda - \lambda^{-2})$ of a rubber elastic network under a uniaxial stress due to the force f. The factor b takes into account the maximum macroscopic strain $\lambda_m = l_{max}/l_0$ determined by the finite chain extensibility. It can be considered to represent an entropy correction factor. The strain factor D_m can be obtained from the maximum elongation λ_m and is equal to the reciprocal of b:

$$D_m = \lambda_m - \lambda_m^{-2} \equiv b^{-1} \tag{285}$$

Furthermore, the maximum elongation λ_m can be derived from the average molar mass of the statistical chain-segment M_s and the average number of such segments

y_s per network chain:

$$\lambda_m = \frac{y_s M_s}{\sqrt{y_s}\, M_s} = \sqrt{y_s} \tag{286}$$

The second parameter a_0 in Eq. 284 describes global interactions between the network chains, i.e. the energy-equivalent subsystems of deformation. With $a = a_0 l_0 / NkT$, Eq.284 can be rewritten as follows:

$$f = \frac{\nu kT}{l_0} D \left[\frac{D_m}{D_m - D} - aD \right] \tag{287}$$

or, since $N = \nu / V$ and $NkT = G$ (see Eq. 242), the stress based on the original cross-section of the sample can be written as:

$$\sigma = GD \left[\frac{D_m}{D_m - D} - aD \right] \tag{288}$$

Experimental stress-strain curves and those calculated on the basis of Eq. 288 for an amorphous SBR and a sample of NR, which crystallizes during elongation, are compared in Fig. 120. The van der Waals' parameters λ_m and a have the same order of magnitude for all known elastomers since the quasi-static deformation behavior of elastomers is essentially independent of the microstructure of their network chains and are thus represented by the properties of a van der Waals conformational gas with weak molecular interactions. Nevertheless, inspite of the doubtless experimental success of the van der Waals theory an all-embracing theoretical description of rubber elasticity, including dynamic processes, requires a more detailed identification of the parameters in microstructural terms.

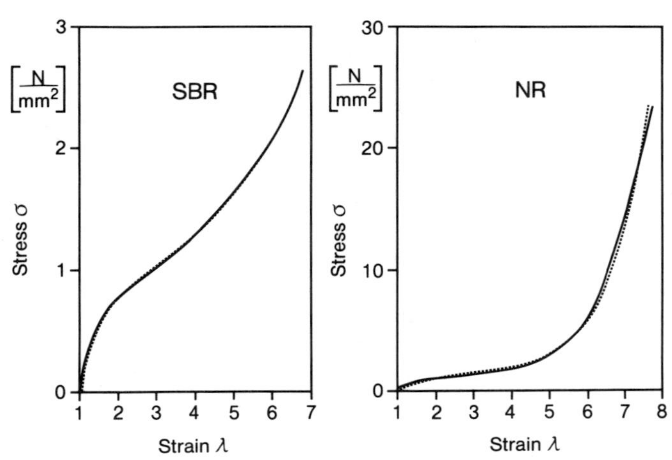

Fig. 120. Uniaxial stress-strain curves for SBR and NR. The *dotted lines* are experimental curves while the *full lines* were calculated on the basis of Eq. 288 with: $\lambda_m = 13.5$ and $a = 0.19$ for SBR and $\lambda_m = 8.8$ and $a = 0.2$ for NR (from [125])

10.11 Photoelastic Properties of Elastomers

Networks and glasses with amorphous structures exhibit isotropic physical properties. The application of a stress to such materials can, however, lead to orientation and a corresponding anisotropy. Highly orientated networks are, for example, optically anisotropic; they exhibit different refractive indices parallel and normal to the applied stress. Such phenomena can be calculated with the following approximations:

a) The network is composed of N chains whose chain vectors r conform to a Gaussian distribution.
b) The deformation is affine.
c) The principal polarizabilities of the complete chain γ_1 and $\gamma_2 = \gamma_3$ can be obtained by integration from the polarizabilities of the individual segments α_1 and α_2
d) An integration over all the chains leads to the total polarizabilities of the network β_1 and β_2 (parallel and normal to the direction of stress respectively).

A partial orientation of N equal, cylindrically symmetrical elements which have the polarizabilities α_1 and α_2 (parallel and normal to their cylinder axes) leads to an anisotropy of the total polarizability:

$$\beta_1 - \beta_2 = N f(\alpha_1 - \alpha_2) \tag{289}$$

In this equation f is an orientation parameter, which, according to P. H. Hermans, can be defined by:

$$f = \frac{1}{2}(3\,\overline{\cos^2\varphi} - 1) = 1 - \frac{3}{2}\,\overline{\sin^2\varphi} \tag{290}$$

whereby φ is the angle between the chain axis and the preferred macroscopic orientation so that f varies between the limits $-1/2 \leqslant f \leqslant 1$.

The Lorentz-Lorenz equation gives the relationship between the refractive index n and the polarizability β:

$$\frac{n^2 - 1}{n^2 + 2} = \frac{4}{3}\pi\beta \tag{291}$$

For small variations $\Delta\beta$ and Δn Eq. 291 can be differentiated to yield:

$$\Delta n = \frac{2}{9}\pi\,\frac{(\bar{n}^2 + 2)^2}{\bar{n}}\,\Delta\beta \tag{292}$$

Substitution of Eq. 289 into Eq. 292 yields the following equation for Δn:

$$\Delta n = \frac{2}{9}\pi\frac{(\bar{n}^2+2)^2}{\bar{n}}Nf(\alpha_1-\alpha_2) \tag{293}$$

The relationship between the elongation of an elastomer and the resulting orientation is:

$$f = \frac{1}{5}(\lambda-\lambda^{-2}) \tag{294}$$

Furthermore, since the stress, according to Eq. 243, is related to the elongation by:

$$\sigma = NkT(\lambda-\lambda^{-2}) \tag{295}$$

it follows that:

$$f = \frac{1}{5}\frac{\sigma}{NkT} \tag{296}$$

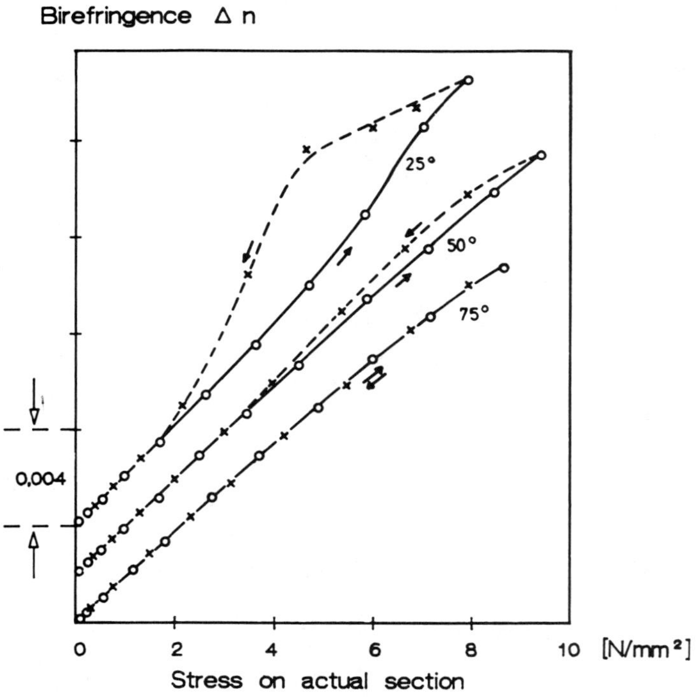

Fig. 121. Birefringence in NR as a function of the applied stress (from [98])

Substitution of Eq. 296 in Eq. 293 yields:

$$n_1 - n_2 = \frac{2\pi}{45kT}\frac{(\bar{n}^2 + 2)^2}{\bar{n}}(\alpha_1 - \alpha_2)\sigma = C_{opt}\sigma \tag{297}$$

where

$$C_{opt} = \frac{2\pi}{45kT}\frac{(\bar{n}^2 + 2)^2}{\bar{n}}(\alpha_1 - \alpha_2) \tag{298}$$

The parameter C_{opt} is called the stress-optical coefficient. The value of this coefficient depends solely on the mean refractive index and the optical anisotropy of the random link. It is independent of the chain length and the degree of crosslinking. From the above it can be seen that birefringence in elastomers is proportional to the applied stress.

In Fig. 121, measurements of birefringence as a function of stress at various temperatures are plotted. These data show a reasonable agreement with linearity as predicted by Eq. 297. The hysteresis loops at lower temperatures can be ascribed to stress-induced crystallization and these phenomena disappear at higher temperatures.

11 Tear Formation and Propagation in Elastomers [130–144]

11.1 Concept of Tearing Energy According to Rivlin

During cyclical deformations of an elastomer small cracks appear on the surface of the sample even though the maximum theoretical stress at peak deformation is much smaller than the tensile strength of the material. Crack growth, the speed of which usually increases with increasing crack length, is often the limiting factor for the useful life of a rubber article. At a defined end of usefulness one speaks of fatigue life. Crack propagation and fatigue life result from stress concentrations at the tips of cracks on the surfaces of rubber articles. These stress concentrations build up at naturally occuring, microscopic surface flaws, extraneous inclusions or inhomogeneities.

On the basis of Griffith's energy criterium, R.S. Rivlin has developed a framework for defining the tear behaviour in elastomers. According to Rivlin's theory, a crack of length c will grow by an infinitely small amount dc only if the energy W, elastically stored in the material, decreases by an amount dW greater than the increase in surface-free energy due to the formation of new surface area.

Thus:

$$-\frac{dW}{dc} > S\,\frac{dA}{dc} \tag{299}$$

where dA is the increase in surface area due to the growth of the crack by dc and S is the specific surface-free energy of the material.

For a sample of thickness d with a tear of length c the quantity S dA = S d dc can be considered to be the amount of work required to produce an increase dc in the cut length. The energy criterium (Eq. 299) can then be written in the form:

$$-\left(\frac{\partial W}{\partial A}\right)_l = S \tag{300}$$

Crack propagation in real elastomers is also influenced by irreversible energy dissipation processes so that an extra term must be added to the energy balance described by Eq. 300:

$$-\left(\frac{\partial W}{\partial A}\right)_l = S + H = T \tag{301}$$

or, with $dA = d\,dc$:

$$-\left(\frac{\partial W}{\partial c}\right)_l = Td \tag{302}$$

In Eq. 302 T is the tearing energy and is characteristic of the material but independent of the specific sample geometry. The subscript l in Eq. 302 indicates that the tear propagation should take place under constant elongation of the sample. As described above, dW is the decrease in the elastic energy stored in the sample. Equation 302 has been solved by Rivlin for a number of different sample geometries.

11.1.1 Trousers Test Piece (see Fig. 122a)

If a trousers sample is pulled in the directions of the arrows shown in Fig. 122a the extension ratio in the shaded area will be λ; the area B remains substantially undeformed in the first instance. When the applied force has increased to a particular level, tear propagation will be initiated. An increase in the length of the cut c by an amount dc reduces the area B. The extension of the material A is unaffected since this depends only on the applied force and this remains constant during tear propagation. The length of the sample (= separation of the clamps) increases, whereby $dl = 2\lambda dc$ when the tear length increases by an amount dc. Since the elastically stored energy W is a function of l and c, one can write:

$$dW = (\partial W/\partial c)_l dc + (\partial W/\partial l)_c dl \tag{303}$$

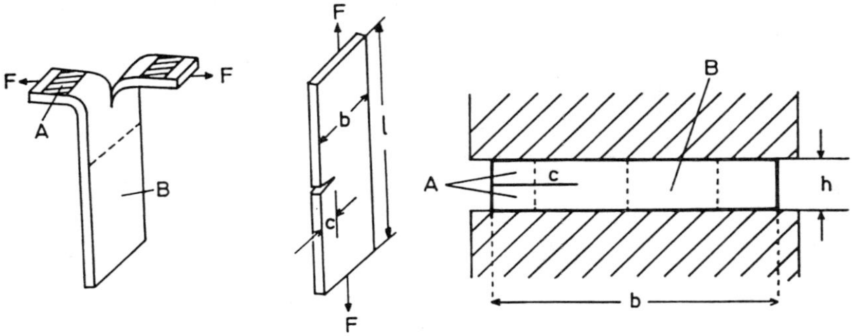

Fig. 122. a) Trousers test piece, **b)** Tensile strip with small cut, **c)** "Pure shear" test piece

and

$$F = (\partial W/\partial l)_c \tag{304}$$

From these equations it is possible to derive:

$$(\partial W/\partial c)_F = (\partial W/\partial c)_l + F(\partial l/\partial c)_F \tag{305}$$

and

$$(\partial W/\partial c)_l = (\partial W/\partial c)_F - 2F\lambda \tag{306}$$

Thus, as the tear length increases by an amount dc a volume of material A_0dc (A_0 is the cross-section of the sample in the area A (Fig. 122a)) is transformed from an undeformed state into a state of simple extension. The amount of elastically stored energy associated with such a transformation: $dW = w_0A_0dc$ (w_o is the energy per unit volume) can be combined with Eq. 306 to give:

$$(\partial W/\partial c)_l = w_0A_0 - 2\lambda F \tag{307}$$

from which, with Eq. 302, one obtains:

$$T = 2\lambda F/d - bw_0 \tag{308}$$

where b is the width of the sample in the region A shown in Fig. 122a.

11.1.2 Tensile Strip with a Small Cut (Fig. 122b)

In order to consider this sample geometry the following assumptions are made: The cut length c is small compared with the width of the test piece, which is, in turn, small compared with its length. Putting the energy stored elastically in the test piece equal to W′ in the absence of a cut and to W in the presence of a cut of length c it can be shown, with the aid of classic elasticity theory, that:

$$\Delta W = W' - W \sim c^2dw_0 \tag{309}$$

where w_0 corresponds to the energy density during a uniaxial elongation (see Eq. 267), i.e.:

$$w_0 = C_1(\lambda^2 + 2/\lambda - 3) + C_2(\lambda^{-2} + 2\lambda - 3) \tag{310}$$

From Eqs. 309 and 310 one can derive:

$$-(\partial W/\partial c)_l = 2kcd\,[C_1(\lambda^2 + 2/\lambda - 3) + C_2(\lambda^{-2} + 2\lambda - 3)] \tag{311}$$

and

$$T = 2kcw_0 \tag{312}$$

Here k is an elongation dependent function which takes account of the excess stress in the area immediately surrounding the tip of the tear. As a first approximation:

$$k(\lambda) = \pi/\sqrt{\lambda} \tag{313}$$

H.W. Greensmith has shown how the function $k(\lambda)$ can be determined experimentally.

11.1.3 "Pure Shear" Test Piece (Fig. 122c)

A more detailed definition of "Pure shear" deformation is shown in Fig. 123.

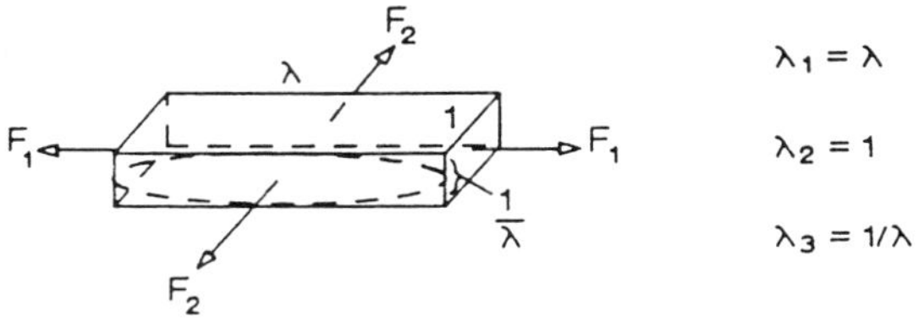

Fig. 123. "Pure shear" deformation

Thus, pure shear is equivalent to a simple shear without superimposed rotation. In terms of the sample in Fig. 122c, this deformation state is realized in the region B if $b \gg h$. As the tear propagates the undeformed region A grows at the expense of region B. The change in the elastically stored energy W corresponding to a tear growth dc is given by:

$$dW = -w_0 hd\,dc \tag{314}$$

from which the tearing energy T can be written in the form:

$$T = w_0 h \tag{315}$$

where w_0 corresponds to the elastic energy density in a pure shear deformation state. The tear propagation energy for this sample geometry is thus independent of the initial length of the cut c.

11.2 Elastic Energy Density in an Elastomer

As has been shown in Sect. 11.1 the critical quantity determining the tear propagation energy of an elastomer is the elastically stored energy w_0. This energy can be easily calculated from the entropy change associated with a deformation (see Eq. 255) and the resulting work $\Delta A = -T\Delta S$. Thus:

$$w_0 = \frac{1}{2} G(\lambda_1^2 + \lambda_2^2 + \lambda_3^2 - 3) \qquad (316)$$

where $G = NkT$ is the shear modulus and λ_1, λ_2 and λ_3 are the principal strains.

For example, for pure shear deformation (see Fig. 123) one obtains from Eq. 316.

$$w_0 = \frac{1}{2} G(\lambda^2 + 1/\lambda^2 - 2) \qquad (317)$$

As a rule however, this equation is inadequate for non-linear deformation. According to Mooney and Rivlin the elastic energy of an incompressible elastomer, in the most general form as a function of the strain invariants I_n (see Eq. 265), can be approximated by:

$$\begin{aligned} w_0 = {} & C_{10}(I_1 - 3) + C_{01}(I_2 - 3) + C_{11}(I_1 - 3)(I_2 - 3) \\ & + C_{20}(I_1 - 3)^2 + C_{02}(I_2 - 3)^2 \end{aligned} \qquad (318)$$

where terms with exponents greater than 2 have been neglected. For a pure shear deformation

$$I_1 = I_2 = \lambda^2 + 1/\lambda^2 + 1 \equiv I \qquad (319)$$

is valid so that:

$$\begin{aligned} w_0 &= (C_{10} + C_{01})(I - 3) + (C_{11} + C_{20} + C_{02})(I - 3)^2 \\ &= \tilde{C}_1[f(\lambda)] + \tilde{C}_2[f(\lambda)]^2 \end{aligned} \qquad (320)$$

Here:

$$f(\lambda) = I - 3 = \lambda^2 + 1/\lambda^2 - 2 \qquad (321)$$

and

$$\tilde{C}_1 = (C_{10} + C_{01}) \quad \text{and} \quad \tilde{C}_2 = (C_{11} + C_{20} + C_{02}) \qquad (322)$$

With Eq. 320 one has an analytical expression for the elastic energy density as a function of the strain invariants and their squares. However, in practice, elastic energy densities are usually obtained experimentally from the integral of the stress-strain curve.

11.3 Fatigue Crack Propagation Under Dynamic Load

Tear initiation and tear propagation in elastomers are phenomena which even occur during cyclic deformation when the peak stress is considerably less than the tensile strength of the sample. In this case, tear growth is accompanied by tearing energies which vary with the amplitude of the deformation. Experiments have shown that this type of tear growth in elastomers can be described by an exponential expression:

$$dc/dn = BT^{\beta} \tag{323}$$

where n is the number of the deformation cycle and B and β are empirical, elastomer specific constants. A comparison of an experimental result with NR and the curve obtained from Eq. 323 with $\beta=2$ is shown in Fig. 124; the agreement is quite acceptable.

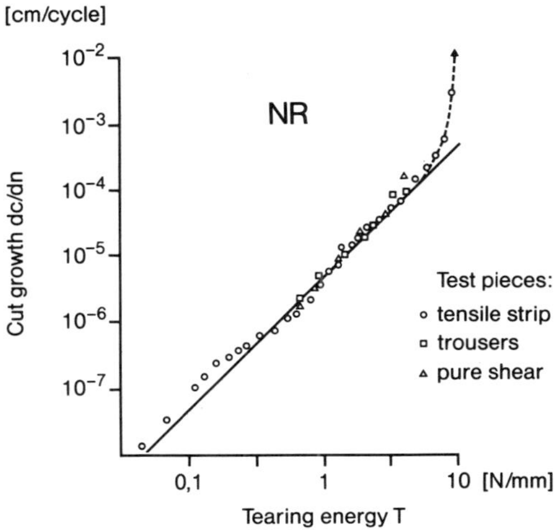

Fig. 124. Cut growth in an unfilled NR sample as a function of the tearing energy: *Full line* is calculated from Eq. 323 with $\beta=2$ (from [137])

D.G. Young has investigated fatigue crack propagation in model vulcanizates of a tire sidewall using a pure shear test piece and pulsed cyclic deformations. His results are reproduced in Fig. 125.

The rate of cut growth in NR compounds is relatively high at low tearing energies (i.e., small strains $<15\%$) but low at higher energies (large strains). The high rates

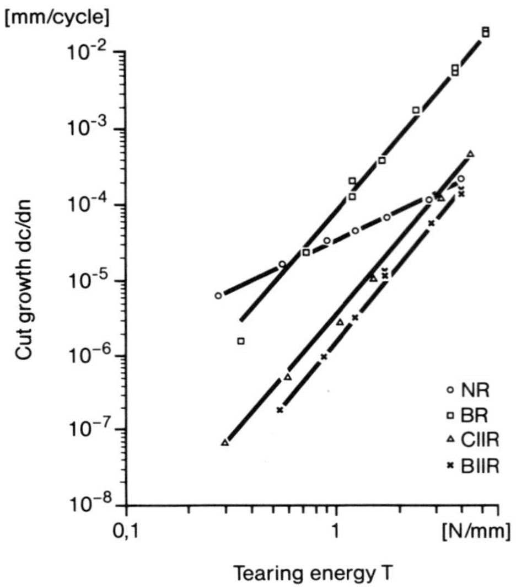

Fig. 125. Fatigue crack propagation in model sidewall compounds as a function of the tearing energy (from [143])

may be explained in terms of the influence of oxidation processes whereas the lower rates are due to the stabilizing effect of strain-induced crystallization.

Using Eq. 323 it is possible to predict the fatigue life of a rubber. Thus, if T in Eq. 323 is substituted by the appropriate term from Eq. 312 one obtains after integration:

$$n = \frac{1}{(\beta - 1)} \frac{1}{B(2kw_o)^\beta} \left[\frac{1}{c_0^{\beta-1}} - \frac{1}{c^{\beta-1}} \right] \tag{324}$$

In this equation c_0 is the original length of the cut. The theoretical course of the cut growth is shown in Fig. 126 in terms of a relative scale and as predicted by Eq. 324.

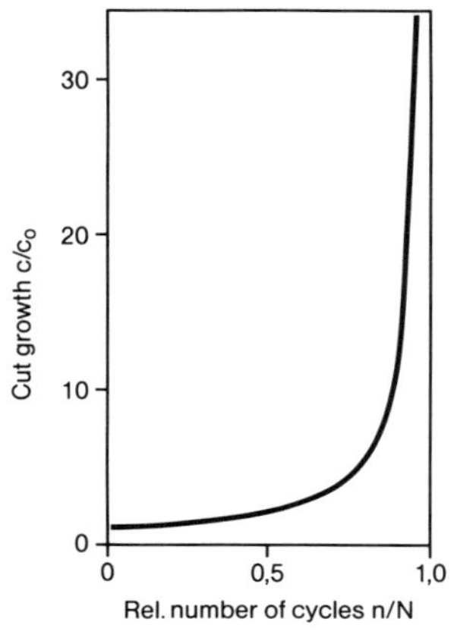

Fig. 126. Cut growth as predicted by Eq. 324 as a function of the relative cycle number with $\beta = 2$

The fatigue life of an elastomer, i.e., the number of deformation cycles it survives before it breaks, can be obtained directly from Eq. 324. That is:

$$N = \frac{1}{(\beta - 1)} \frac{1}{B(2kw_0)^\beta} \frac{1}{c_0^{\beta - 1}} \tag{325}$$

which takes into account that at break $c \gg c_0$.

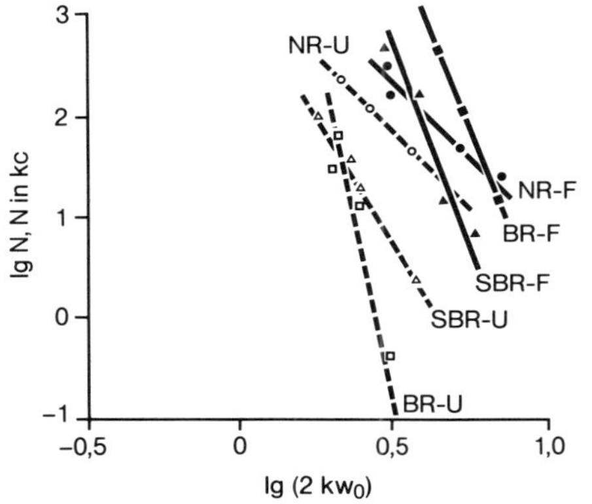

Fig. 127. Fatigue life of several filled (F) and unfilled (U) elastomers (f = 100/min) (from [139])

The results of experiments on the fatigue life of several unfilled and filled elastomers in terms of Eq. 325 are shown in Fig. 127. From the slopes of these plots the elastomer specific constants β can be obtained. There are often considerable differences between the values of β obtained from tear propagation and fatigue life experiments. This reflects the fact that tear propagation in an elastomer during dynamic deformation cannot be adequately described by Eq. 323 over the whole range of tearing energies.

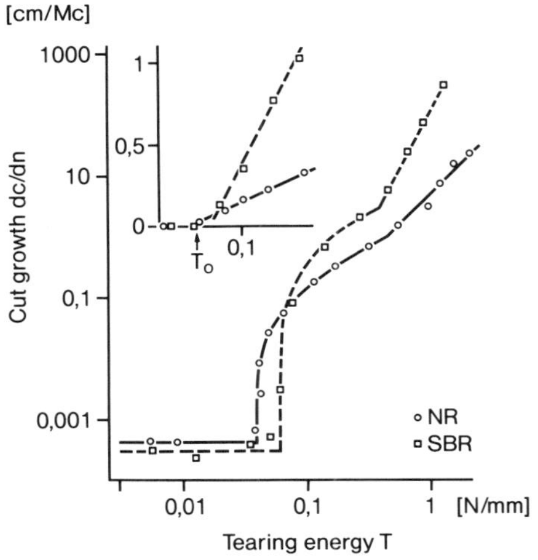

Fig. 128. Cut growth in unfilled NR and SBR in the region of low tearing energy. *Insert:* Linear scale in the region of T_0 (from [144])

Several authors have reported experiments which suggest that the course of cut growth as a function of tearing energy has the form shown in Fig. 128. Obviously, there are several regions where the rate of tear growth varies. Below a critical energy T_0 (approx. 0.05 N/mm in Fig. 128) the speed of tear growth in elastomers is essentially constant and limited by the concentration of ozone in the environment. In vacuo no tear growth occurs below T_0. The critical tearing energy corresponds to a critical strain which lies between 50 and 100% for most synthetic elastomers. The rate of cut growth in the region just above T_0 is considerably increased by the action of oxygen and this type of growth is therefore referred to as mechanico-oxidative. T_0 thus represents the minimum tearing energy at which mechanico-oxidative cut growth can occur.

In principle, the value of T_0 is dependent on the strength of the chemical bonds in the polymer chain and an approximate calculation on this basis has indeed been made.

The mechanico-oxidative tear growth is described adequately by the exponential Eq. 323. It should be noted that, since the exponent β is a polymer specific parameter, a double logarithmic plot according to this equation leads to characteristic straight lines for different elastomers and that these cross each other. Analogous intersections occur for the fatigue life curves shown in Fig. 127. This phenomenon has practical consequences in terms of the fatigue resistance of elastomeric materials under the influence of both chemical and mechanical stresses.

As already indicated in Fig. 124, after reaching a critical tearing energy limit catastrophic, rapid tear growth occurs. Fig. 129 shows the dependence of this limiting energy T_c (as measured in quasi-static stress-strain experiments on samples similar to those shown in Fig. 122) as a function of the degree of crosslinking.

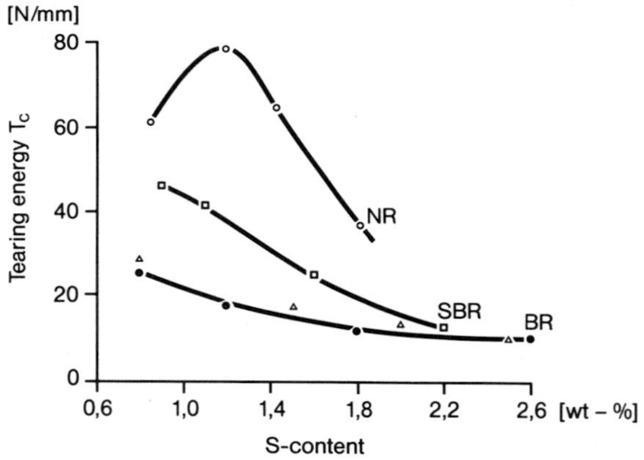

Fig. 129. Critical tearing energy T_c of catastrophic tear growth as a function of the degree of crosslinking for several elastomers.

From Fig. 129 the excellence of natural rubber in comparison to, for example, the synthetic elastomers SBR and BR is very obvious. Thus, NR is often preferred for use in more demanding rubber articles.

12 Deformation Behavior of Thermoplastic Elastomers [145–157]

12.1 Structural Principles

Entropy elastic deformation of polymers requires that the molecular chains cannot dissipate an applied strain via plastic flow. This state is achieved in covalently crosslinked elastomers by statistically bonding neighboring chains together, for example, via S-bridges. An alternative way of hindering plastic flow is to introduce moieties which form strong physical bonds between the chains into the network. The two principal types of network are shown schematically in Fig. 130. In the physical network the parts of the molecules which are enclosed by the rings shown in Fig. 130 are held together by non-covalent interactions so that these regions form effective, large volume crosslinks. These regions, together with the entropy elastic chains outside these physical crosslinks, allow such materials to exhibit considerable elasticity.

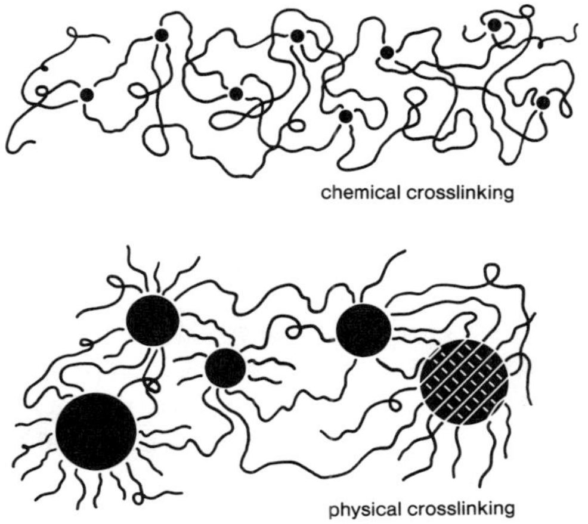

Fig. 130. Sketch comparing chemically and physically crosslinked networks

12.2 Polyurethane Elastomers

The principle of physical crosslinks is realized by a technologically important group of polymers – the polyurethane elastomers (PUR). The PUR chains, as shown in Fig. 131, are constructed from flexible and stiff blocks.

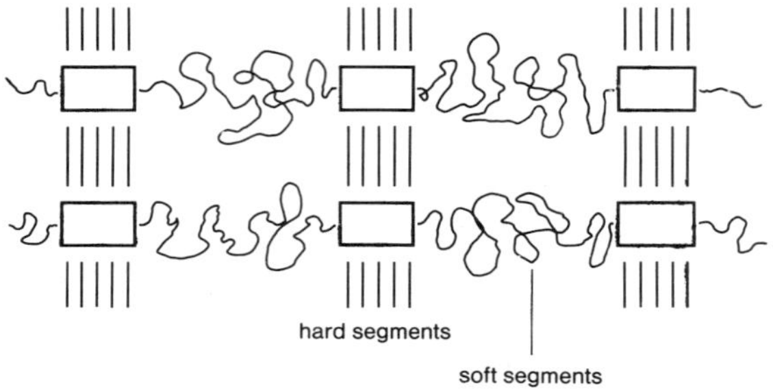

Fig. 131. Physical cross-linking in PUR-elastomers

Thermoplastic polyurethanes are synthesized essentially from three components:

a) long chain diols HO~~~~OH
b) short chain diols HO-R-OH
   or diamines H_2N-R-NH_2
c) diisocyanates OCN-R-NCO

The reaction of the three components begins at temperatures between 60 and 140 °C with rigorous stirring (one shot process). During the reaction the isocyanate and hydroxy functions react to form urethane groups without the production of any by-products: This is a polyaddition reaction. As the reaction progresses two phases start to develop and the separation continues as the material cools after the reaction is complete. The soft phase contains the long-chain diols which can also have been chain-extended by single diisocyanate molecules. The hard phase is composed of polyurethane blocks formed from diisocyanates and chain extender (short chain diol or diamine). The hard domains are ordered regions which are held together by hydrogen bonding.

Physical crosslinking has several advantages over chemical crosslinking: Thus, for example, at temperatures of about 200 °C the hard segments in PUR elastomers become plastic and these materials can be processed as thermoplastics (e.g., by injection molding). At lower temperatures the PUR-elastomers exhibit rubber elasticity since the soft segments have glass temperatures of about –40 °C. PUR-elastomers, from softer rubbers to impact resistant plastics, can be synthesized simply by varying the amounts of hard and soft segments. The excellent tensile properties of PUR-elastomers result from the relatively uniform crosslinking. Furthermore, the reorganisation of the hard segments during deformation allows applied stresses to be spread over an optimum number of covalent bonds (Fig. 132).

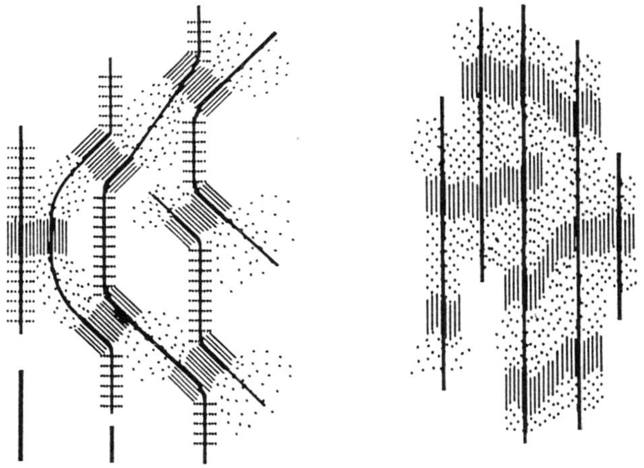

Fig. 132. Homogenization of stress in PUR-elastomers (from [147])

The reorganization, as a consequence of sliding within the hard segments, thus allows the stress to be homogenized without breaking any chains. This process leads to an orientation of the hard segments and, in the extreme, to stress induced crystallization of the soft segments.

The polyurethanes are an extremely versatile group of materials: In addition to elastomer articles, they also find use in the form of fibers, as glues, paints and even foams. For the manufacture of foams the reaction of isocyanates with water, which yields carbon dioxide as a side product, is exploited. Thus, if water is added to a polyol-diisocyanate mixture the isocyanates can either react with a hydroxy group or with water. As the reaction proceeds and the polyurethane macromolecules are built up the increasingly viscous mass is expanded by myriad bubbles of carbon dioxide.

12.3 Block Copolymers

The most well known of this class of materials are the linear or radial styrene-butadiene-styrene block copolymers. In these materials the physical crosslinks are produced by the polystyrene domains which are in a glassy state at room temperature. Figure 133 is a sketch showing the phase structure of a SBS triblock copolymer. The form and size of the domains in such materials depend, among other parameters, on the volume fraction of the individual components and their molar masses. Since the driving force for the production of a particular morphology is a reduction of interfacial area, spherical, cylindrical and layer morphologies can be prepared by varying the amounts of the components (see Fig. 134).

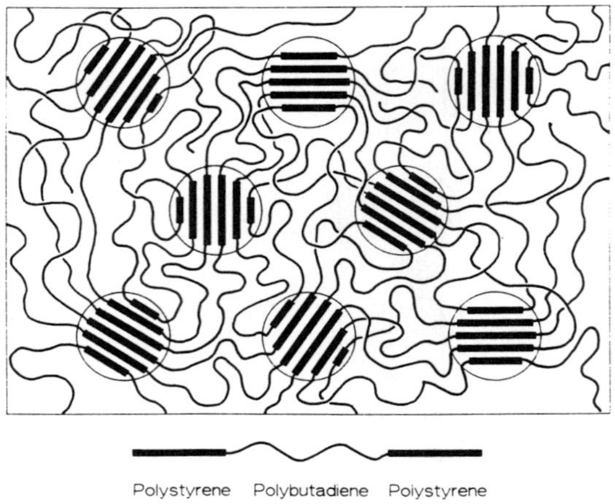

Fig. 133. Domain structure in an SBS triblock copolymer

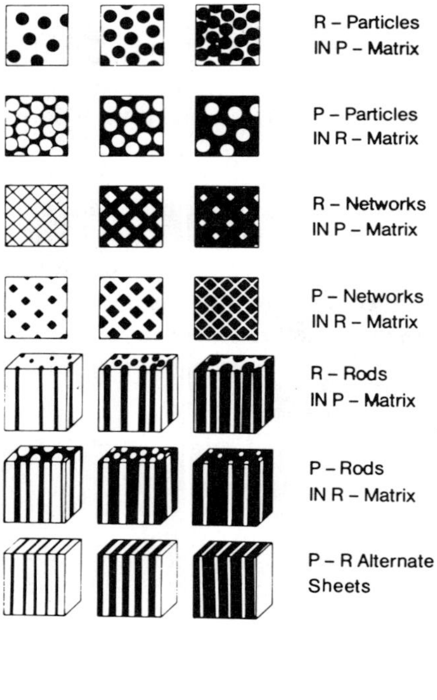

Fig. 134. Sketch showing the regular, supramolecular structures in a two-phase system of components "R" and "P" (from [152])

The availability of technologically viable processes for synthesizing block copolymers with well-defined compositions has led to many of these regular, supramolecular structures being commercially realized. As an example, Fig. 135 shows an electronmicrograph of a SB diblock system which has well-defined, cylindrical domains of butadiene regularly ordered in a coherent polystyrene matrix.

Fig. 135. Electronmicrograph of an SB diblock copolymer (68 vol-% styrene) (from [153])

Because of their homogeneous network structure, such block copolymers exhibit exceptionally good tensile properties equivalent to those of covalently crosslinked elastomers reinforced with fillers. The two-phase morphology of these block copolymers also manifests itself in two glass transitions corresponding to those of the two components (see Fig. 136).

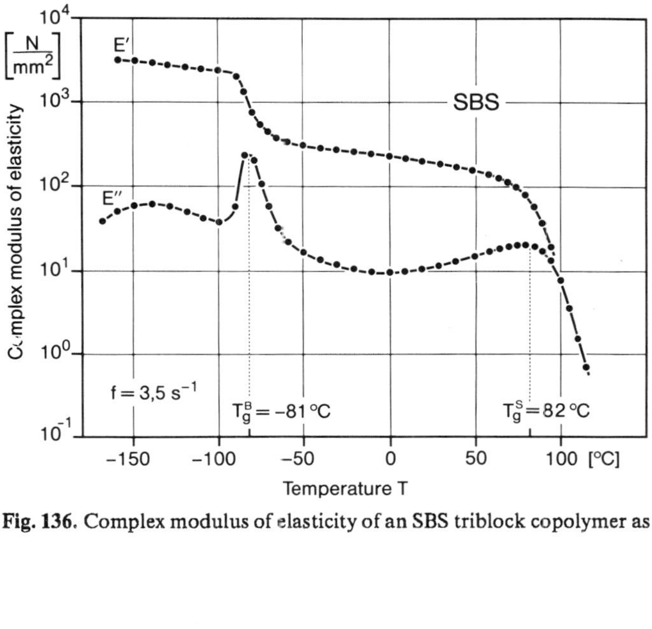

Fig. 136. Complex modulus of elasticity of an SBS triblock copolymer as a function of temperature

The temperature range in which these materials exhibit rubber elasticity is limited by the glass transitions of their two components. Thus, application of such materials over a wide temperature range requires that the glass transitions of the elastomer and the thermoplast are as low and as high as possible, respectively. As can be seen in Fig. 136, the polystyrene in SBS-systems softens at rather low temperatures so that these materials are not suitable for use at temperatures above about 70 °C. Modern technologies require materials with form stability at increasingly higher temperatures and considerable industrial research is aimed at substituting the thermoplast phase in thermoplastic elastomers to achieve better temperature stability. To this end, increasing use is being made of partially crystalline materials which, in part due to their relatively high melting temperatures, provide better high temperature properties.

12.4 Thermoplastic Elastomers Based on Polymer Mixtures

Thermoplastic elastomers can also be prepared simply by mixing two polymers. The most important criterium for the success of such mixtures is the degree of incompatibility of the two polymers which leads to the development of a multiphase morphology. One commercially successful system based on this principle is a mixture of isotactic polypropylene and statistical or sequential ethylene-propylene copolymers (EPM or EPDM, the latter being a tercopolymer with a diene). In such mixtures there are, formally, no covalent bonds between the hard and soft phases. One way of looking at such systems is to consider that a coupling of the two phases through entanglement of amorphous polypropylene and EPM molecules occurs. Since the hard phase is composed of crystalline polypropylene, such materials retain their form even at temperatures well above 100 °C (T_m (Polypropylene) = 160 °C). Modulus/temperature curves for a range of PP/EPDM mixtures are shown in Fig. 137. For all the mixtures two glass transitions can be identified (for PP at +3 °C; for EPDM at −37 °C) indicating that the mixtures have two separate phases. Noteworthy, is the continuous decrease in the modulus over the whole temperature range. A reduction in this undesirable characteristic can be attained by crosslinking the materials during mixing with e.g. peroxides. Such dynamic, in situ vulcanization can lead to technologically interesting morphologies e.g., essentially fully cross-linked rubber particles dispersed in the thermoplast matrix.

One disadvantage of this class of thermoplastic elastomers are their relatively poor tensile properties due to the relatively weak bonds between the hard and soft phases compared with those of the block systems. The coupling of the two phases can be improved by the addition of a third, compatibilizing component. In recent years, graft copolymers have been added to polymer mixtures to this end. With an appropriate choice of graft component, such that this is either identical or at least compatible with the matrix polymer, the coupling of the dispersed phase onto the matrix can be considerably improved.

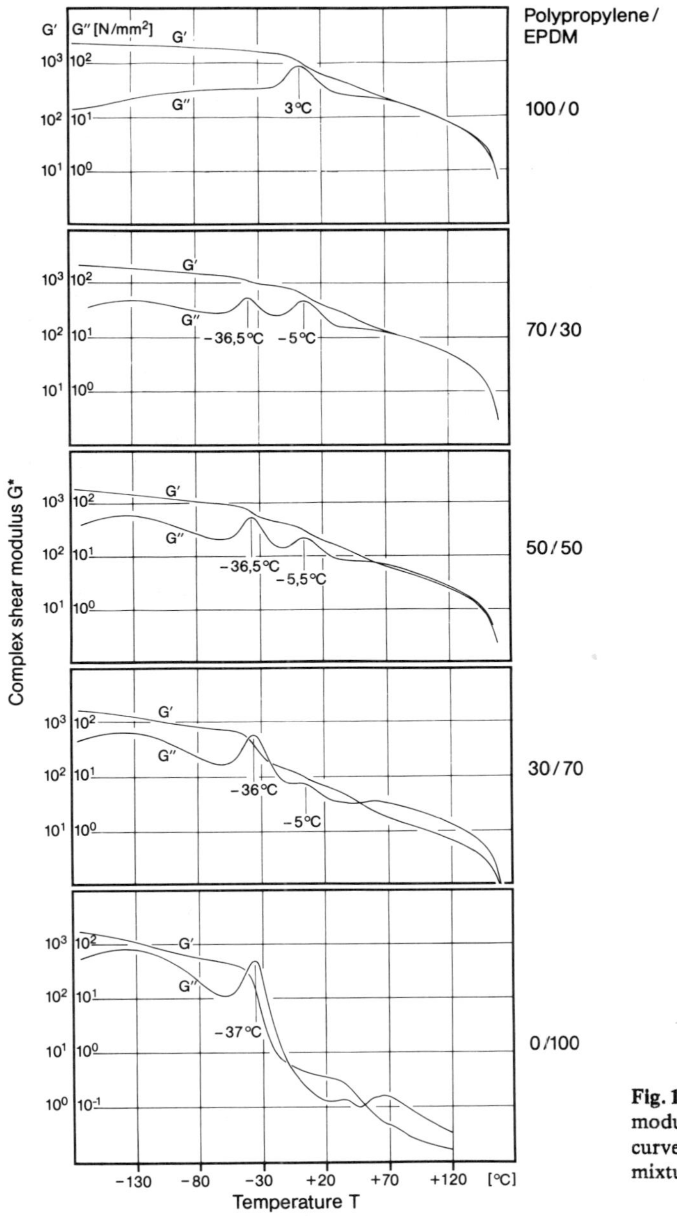

Fig. 137. Complex shear modulus/temperature curves for some EPDM/PP mixtures

Some commercial importance has been achieved by multi-component systems in which two or more phases form a co-continuous morphology, the socalled interpenetrating networks (IPN).

In principle, the ordering of the individual phases and the interactions between these determine the tensile properties of thermoplastic elastomers.

12.5 Tension Set

The question of the possibility of substituting all classically crosslinked elastomers by thermoplastic elastomers is frequently discussed. Elastomers with permanent, covalent crosslinks are characterized by their deformation being, to a high degree, reversible. In this respect they are considerably better than any other class of materials. A particularly suitable experiment to quantify this behavior is the intermittent stress-strain experiment. Figure 138 shows the result of such experiments on an elastomer with covalent crosslinks, a thermoplastic elastomer and a thermoplast. In these experiments a sample is stretched to a predetermined elongation, released and then stretched to the next greater degree of elongation and so on. The remaining strain at a given time after the removal of the applied stress, the tension set, is a measure of the irreversibility of the deformation. The tension sets derived from Fig. 138 as a function of the predetermined elongation are plotted in Fig. 139. The dotted line in Fig. 139 represents the line of totally irreversible deformation, the abscissa totally reversible deformation.

It can be seen from Fig. 139 that the deformation of a thermoplast is almost totally irreversible whereas that for an unfilled NR vulcanisate is almost totally reversible. In principle, it should be possible to draw limits on such a diagram

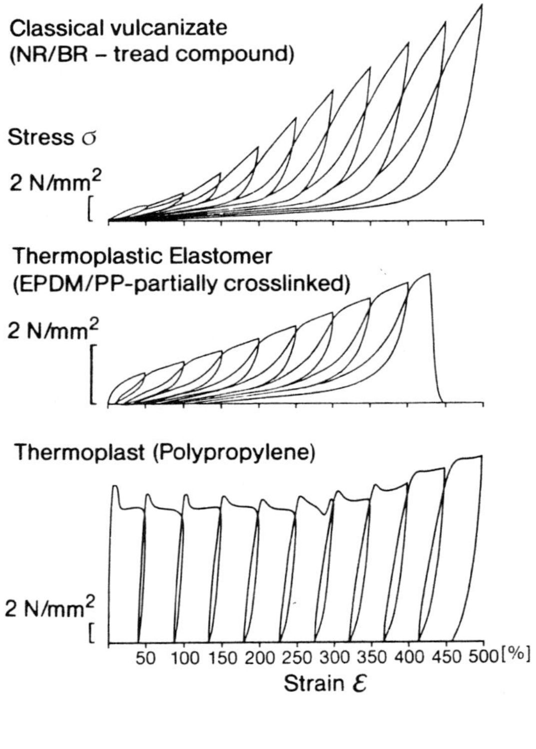

Fig. 138. Experimentally measured curves from intermittent stress-strain experiments on an elastomer with covalent crosslinks (*top*), a thermoplastic elastomer (*middle*) and a thermoplast (*bottom*)

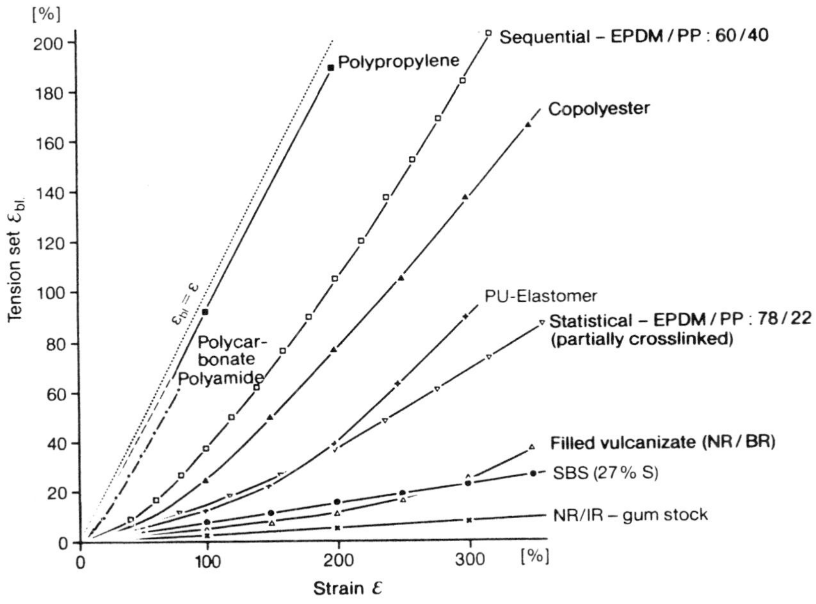

Fig. 139. Tension sets ε_{bl} of different polymers

between the different types of polymers i.e., those which are essentially elastomeric and those which are essentially thermoplastic in character. However, the modulus of the thermoplastic elastomers or elastomer modified thermoplastics in use needs to be considered, since a high degree of reversible deformation and a high modulus, or indeed the inverse are not mutually exclusive. Most thermoplastic elastomers have considerably greater tension sets than covalent elastomers, even when the latter are highly filled. An exception is the reversibility of deformation of SBS-triblock copolymers. However, the tension set of all thermoplastic elastomers, as well as SBS, increases rapidly as the temperature is increased. In contrast, the tension sets of covalently crosslinked elastomers remains constant over a wide temperature range. The greater degree of irreversible deformation exhibited by thermoplastic elastomers can be regarded as a severe limit to their general substitution for classical elastomers. Nevertheless, it should be emphasized that in view of their interesting property profiles and from an economic point of view (e.g. dispensing with the cost and energy intensive vulcanization process) thermoplastic elastomers must be accorded an important future.

Part IV:
Mixing and Swelling of Polymers

13 Compatibility of Polymers [158–190]

13.1 Basic Theoretical Considerations

Mixtures of polymers do not, generally, form thermodynamically stable single-phase systems. Complete miscibility in a mixture of two polymers requires that the free energy of mixing is negative:

$$\Delta G_m = \Delta H_m - T\Delta S_m < 0 \tag{326}$$

The enthalpy ΔH_m and entropy ΔS_m of mixing are generally both positive for pairs of polymers so that only if

$$\Delta H_m < T\Delta S_m \tag{327}$$

will the two polymers mix to form a single phase. Equation 327 defines a lower limit for ΔH_m above which, two polymers are not miscible. An additional, necessary and sufficient condition for compatibility is given by:

$$\left[\frac{\partial^2 \Delta G_m}{\partial \phi_i^2}\right]_{T,p} > 0 \tag{328}$$

where ϕ_i is the volume fraction of the ith component. Three possible forms for binary systems are shown in Fig. 140. Curve B corresponds to a system of two totally miscible polymers. Curve A is that corresponding to a totally incompatible polymer

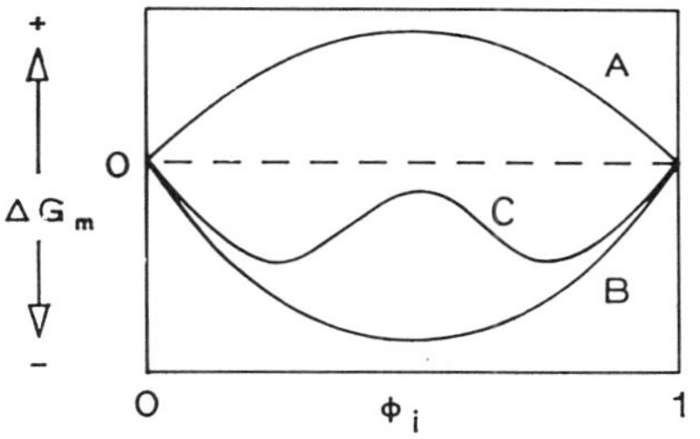

Fig. 140. Free energy of mixing for binary mixtures as a function of their composition: *A)* Incompatibility; *B)* Total miscibility; *C)* Partial miscibility

system and curve C represents a partially miscible system. In addition, curve C exhibits a gap in miscibility where conditions laid down by Eq. 328 are not fulfilled.

The behavior of binary systems can best be expressed by their phase diagrams, i.e. plots showing the relative compositions of the mixtures as a function of temperature. Two examples of the temperature dependence of mixing phenomena are shown in Fig. 141a/b and 142a/b. These examples emphasize that phase separation can occur both due to an increase in temperature (Fig. 141a/b) or due to

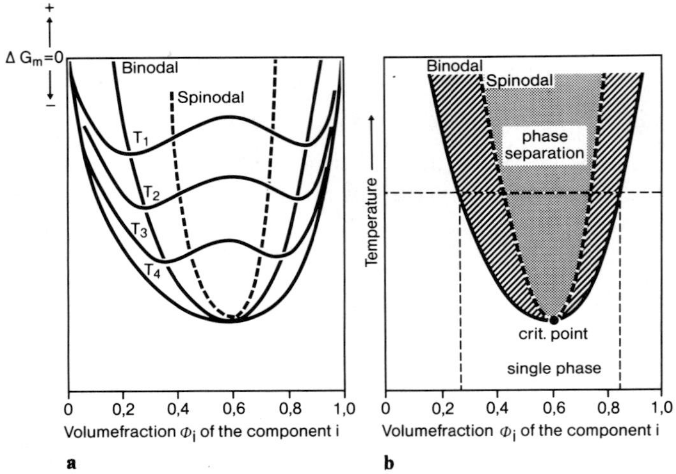

Fig. 141. **a** Free energy of mixing as a function of the relative composition of a binary mixture; $T_1>T_2>T_3>T_4$. **b** Coexistence curve for a binary mixture with an upper miscibility gap

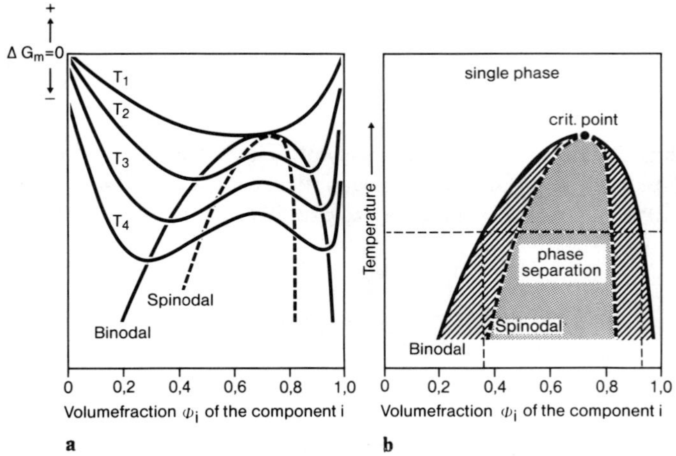

Fig. 142. **a** Free energy of mixing as a function of the relative composition of a binary mixture; $T_1>T_2>T_3>T_4$. **b** Coexistence curve for a binary mixture with a lower miscibility gap

a decrease in temperature (Fig. 142a/b). The area in such a diagram where phase separation occurs is called the *upper* (Fig. 141) or the *lower* (Fig. 142) miscibility gap. For any given temperature the two points on the boundary curve (the binodal) for single phase behavior can be extrapolated to the horizontal axis to give the composition of the two mixed phases which form on phase separation. Neither of these two phases is pure but consists predominantly of one component with a small amount of the second dissolved in it. One point on the binodal is rather special: The critical separation point where both mixed phases have the same composition. Mixed phase and miscibility gap are not distinguishable at this point. The curves joining minima or points of inflection are called the binodal or spinodal respectively. In the shaded area between the binodal and the spinodal the system is metastable.

13.2 Flory-Huggins Theory [163, 164]

Both Flory and Huggins have attempted, independently, to describe the mixing behavior of polymers in low molar mass solvents. As a solvent molecule diffuses into a polymer the entropy of the total system increases (entropy of mixing) while the internal energy of the system remains essentially constant. A simple approach to calculating the entropy of mixing has been given by Flory: The solvent molecules are assumed to be ordered as a 3-D lattice, whereby each lattice point can be occupied by a solvent molecule or a single polymer segment.

Putting:

N: Total number of lattice sites
n: Number of polymer molecules with x segments and
m $= N - xn$: Number of solvent molecules.

The problem involves calculating the number of different possible configurations of the n polymer molecules and $N - xn$ solvent molecules. As a start, the individual segments of the polymer molecules are arranged in the lattice, whereby consecutive polymer segments have to be sited at neighboring lattice sites. The second step is to fill the vacant lattice sites with solvent molecules. Thus, with n_i polymer molecules in the lattice, the first segment of the $(i+1)$ molecule can be put in any one of the remaining $N - xn_i$ vacant lattice sites. The positioning of the second segment of this molecule is restricted to those free lattice sites immediately adjacent to the site occupied by the first segment. The fraction of the lattice sites not occupied by polymer segments is given by $(N - xn_i)/N$. Putting the number of sites adjacent to that occupied by the first segment to z, the number of available lattice sites for the second segment is given by: $z(N - xn_i)/N$. By analogy, one can derive the expression for the available lattice sites for the third segment of the polymer chain to be:

$(z-1)(N-xn_i)/N$. Flory then uses the approximation that, for the fourth and remaining polymer segments, an equal number of lattice sites are available:

$$y = (z-1)\frac{(N - xn_i)}{N} \tag{329}$$

The total number of configurations for the individual polymer molecules is then given by:

$$v_{i+1} = \frac{(N - xn_i)}{2}\,\frac{zy^{x-1}}{z-1} \approx \frac{(N - xn_i)^x}{2}\left[\frac{z-1}{N}\right]^{x-1} \tag{330}$$

The factor 1/2 in this equation takes account of the possibility of choosing either of the two polymer chain ends as a starting point for arranging of the polymer molecule in the lattice. The total number of different configurations for an ensemble of n polymer molecules in a lattice is given by:

$$W = \frac{1}{n!}\prod_{i=1}^{n} v_i \tag{331}$$

The configurational entropy can be obtained from Eq. 331 with Eq. 330 and the Boltzmann equation, $S = k \ln W$:

$$S = -k\left[m \ln \frac{m}{(m + xn)} + n \ln \frac{n}{(m + xn)}\right] + k(x-1)n[\ln(z-1) - 1] - kn \ln 2 \tag{332}$$

The entropy of mixing ΔS_m is given by the difference between S (as given by Eq. 332) and the configurational entropy for the undissolved polymer. The latter can be obtained from Eq. 332 with $m = 0$.

Thus:

$$\Delta S_m = -k\left[m \ln \frac{m}{m + xn} + n \ln \frac{xn}{m + xn}\right] \tag{333}$$

or since $m + xn = N$, Eq. 333 can also be written in the form:

$$\Delta S_m = -k(m \ln \phi_m + n \ln \phi_n) \tag{334}$$

In this equation ϕ_n and ϕ_m are the volume fractions of the polymer and the solvent, respectively.

Differentiation of Eq. 334 with respect to m gives the molar entropy of mixing:

$$\Delta s_m = -R[\ln(1 - \phi_n) + (1 - 1/x)\phi_n] \tag{335}$$

A calculation of the free energy of mixing requires a knowledge of the enthalpy of mixing. An exact calculation of the latter including all the interactions in a mixed system is not possible since it is a multi-particle problem with a large number of potential parameters. For this reason, as a first approximation, only the interactions between adjacent solvent molecules and polymer segments are taken into account. In terms of the enthalpy of mixing, only the difference between the total energy of interaction in the solution and the energy of interaction of the pure components is relevant. Putting the interaction between similar entities to w_{11} or w_{22} and those between different components to w_{12}, the energy change on contact between two components can be written in the form:

$$\Delta w_{12} = w_{12} - \frac{1}{2}(w_{11} + w_{22}) \tag{336}$$

On formation of a total of p_{12} contact pairs in solution, the enthalpy of mixing is given by:

$$\Delta H_m = p_{12}\Delta w_{12} \tag{337}$$

The probability that a lattice site adjacent to a given polymer segment is occupied by a solvent molecule is assumed to be proportional to the volume of solvent present. The number of possible contacts per polymer molecule consisting of x segments is equal to zx. From this one obtains for the enthalpy of mixing:

$$\Delta H_m = zxn\phi_m\Delta w_{12} = zm\phi_n\Delta w_{12} \tag{338}$$

With

$$\chi = \frac{z\Delta w_{12}}{kT} \tag{339}$$

Eq. 338 can be rewritten in the form:

$$\Delta H_m = kTm\chi\phi_n \tag{340}$$

The dimensionless parameter represented by χ and called the interaction parameter, is a measure of the energy of interaction per solvent molecule divided by kT. With Eq. 334 and 340 the free energy of mixing is given by:

$$\Delta G_m = \Delta H_m - T\Delta S_m = kT(m \ln \phi_m + n \ln \phi_n + \chi m\phi_n) \tag{341}$$

and (using Eq. 335) in molar terms:

$$\Delta g_m = RT[\ln (1 - \phi_n) + (1 - 1/x)\phi_n + \chi\phi_n^2] \tag{342}$$

Equation 342 is known as the Flory-Huggins equation. If the number of chain segments $x \gg 1$, then the term $1/x$ can be neglected and Eq. 342 can be simplified to:

$$\Delta g_m = RT[\ln(1 - \phi_n) + \phi_n + \chi\phi_n^2] \qquad (343)$$

13.3 Development of the Flory-Huggins Theory to a Description of Polymer Mixtures

The lattice theory for the enthalpy of mixing in polymer solutions, developed by Flory and Huggins, can be applied formally to mixtures of polymers. In analogy to Eq. 334, the entropy of mixing for two polymers is given by:

$$\Delta S_m = -k[n_1 \ln \phi_1 + n_2 \ln \phi_2] \qquad (344)$$

The enthalpy of mixing can be written, analog to Eq. 340:

$$\Delta H_m = kT\chi_{12}N\phi_1\phi_2 \qquad (345)$$

where ϕ_i is the volume fraction of the polymer i and $N = n_1 + n_2$ is the total number of polymer molecules in the mixture. Without wanting to detract from the importance of the Flory-Huggins theory, it must be stated that it allows only a rough estimate of the miscibility of polymer systems. For example, neither reliable data on the basis of intra- and intermolecular potentials for the empirical interaction parameter χ are available, nor is it realistic to reduce the mixture to the possibilities of the molecules occupying a simple, stiff lattice as this theory does. As a consequence, the Flory-Huggins model does not take account of volume effects which play an important role in the mixing of polymers. More complex theories, such as the equation of state theories (Flory-Prigogine, Sanchez et al.),have been developed from the equations of state for the individual components and do allow a more or less complete description of the p-V-T behavior of polymer mixtures. However, these theories have the disadvantage that they do not, ab initio, take account of the chemical structure of the polymer molecules but rather of the interactions, also in terms of an empirical interaction parameter.

13.4 Solubility Parameter

Due to the extreme difficulties associated with an exact ab initio description of the miscibility of polymers, semi-empirical approaches which allow practical predictions to be made, are of considerable interest. The solubility parameter, with which the practical chemist can estimate the miscibility of polymers, is one such approach. The initial assumption of this approach is that amorphous polymers can be treated as liquids at temperatures above their glass transitions. According to Hildebrand, the molar cohesion energy of a liquid E_K can be defined as the energy required to cancel all intermolecular interactions in one Mole of the liquid. Furthermore, this quantity of energy corresponds to the internal molar enthalpy of evaporation and is related to the solubility parameter δ in terms of:

$$\delta = (E_K/V)^{1/2} \tag{346}$$

The term E_K/V is called the cohesion energy density. The enthalpy of mixing per unit volume can then be written in terms of the solubility parameter as:

$$\frac{\Delta H_m}{V} = \phi_1 \phi_2 (\delta_1 - \delta_2)^2 \tag{347}$$

For $\delta_1 = \delta_2$ it follows that $\Delta H_m = 0$. Thus, miscibility results due to the negative entropy of mixing according to Eq. 326. As the difference between the solubility parameters increases, the degree of miscibility decreases.

The cohesion energy, and thus the solubility parameters, can be separated into three contributions:

$$\delta^2 = \delta_d^2 + \delta_p^2 + \delta_h^2 \tag{348}$$

where δ_d^2: Contribution from dispersion forces, δ_p^2: Contribution from polar interactions δ_h^2: Contribution from hydrogen bonding.

It is possible to calculate the solubility parameter of a polymer using a group additivity approach so that estimates for the miscibility of polymers can be made. One suitable, additive quantity is the molar attraction constant F_i, proposed by Small and defined by:

$$F_i = (E_i v_i)^{1/2} \tag{349}$$

where E_i and v_i are the cohesion energy and the molar volume of the group being considered.

From the additivity of the F_i-values and Eq. 346 one obtains:

$$\delta = \varrho \frac{\sum_i F_i}{M} \tag{350}$$

In Table 8 some values of F_i, from which solubility parameters can be calculated with the aid of Eq. 350, are listed.

Table 8. Group Contributions to F_i (from [27])

Group	F_i [$J^{1/2}$ $cm^{3/2}$/mol]		
	Small [169]	van Krevelen [27]	Hoy [170]
$-CH_3$	438	420	303.4
$-CH_2-$	272	280	269
$-CH(CH_3)-$	495	560	(479.4)
$-CH{=}CH-$	454	444	497.4
$-C(CH_3){=}CH-$	(704)	724	(724.9)
Phenyl	1504	1517	1398.4
$-Cl$	552	471	419.6
$-CN$	839	982	725.5
$-OH$		754	462
$-O-$	143	256	235.3
$-CO-$	563	685	538.1
$-COO-$	634	512	668.2
$-S-$	460	460	428.4

A comparison between some experimentally determined and calculated solubility parameters is made in Table 9.

Table 9. Experimentally determined and Calculated Solubility Parameters for Several Polymers (From [27])

Polymer	δ_{exp} [$J^{1/2} cm^{-3/2}$]	δ_{calc} [$J^{1/2}$ $cm^{-3/2}$]
Polyethylene	15.8–17.1	16.0
Polypropylene	16.8–18.8	17.0
Polystyrene	17.4–19.0	19.1
PVC	19.2–22.1	19.7
Polymethylacrylate	19.9–21.3	19.9
PMMA	18.6–26.2	19.0
Polyacrylonitrile	25.6–31.5	25.7
Polybutadiene	16.6–17.6	17.5
Polyisoprene	16.2–20.5	17.4
Polychloroprene	16.8–18.9	19.2

Solubility and interaction parameters for polymers of similar size are dependent in the following way: According to Eq. 340 the molar enthalpy of mixing for a binary system is given by:

$$\Delta h_m = RT\chi_{12}\phi_1\phi_2 \tag{351}$$

The same quantity can, according to Eq. 347, be written as:

$$\Delta h_m = \phi_1\phi_2 v(\delta_1^2 - \delta_2^2) \tag{352}$$

with v as the molar volume. A combination of these equations leads to the relation between the interaction and the solubility parameters:

$$\chi = \frac{v}{RT}(\delta_1^2 - \delta_2^2) \tag{353}$$

In terms of molar mass there is a critical upper limit for the interaction parameter above which two polymers become immiscible. This critical value can be determined from:

$$(\chi_{12})_{crit} = \frac{1}{2}\left[\frac{1}{P_1^{1/2}} + \frac{1}{P_2^{1/2}}\right] \tag{354}$$

Here, P_i represents the degree of polymerization for the ith polymer component. For $\chi_{12} < (\chi_{12})_{crit}$ the two polymers 1 and 2 are compatible. With increasing molar mass the miscibility of any polymer pair decreases.

13.5 Experimental Methods for Determining Miscibility

The measurement of the complex modulus as a function of temperature is the most important method for determining the miscibility and the phase behavior of polymer mixtures. In Fig. 143, four possible cases are shown schematically. *Case 1* shows a curve (*solid line*) indicating two glass transitions corresponding to those of the individual components A and B (*broken lines*). The two polymers are completely immiscible and are present as two separate phases. In contrast, the curve in *case 2* indicates one sharp glass transition roughly in between those of the two compo-

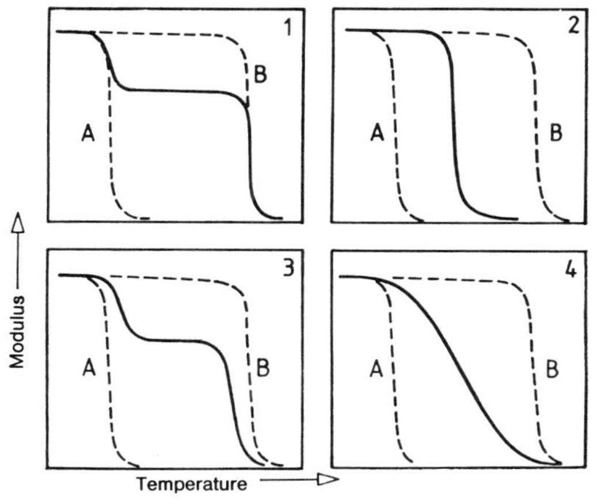

Fig. 143. Principal possibilities for the modulus/temperature behavior of a two phase polymer system

nents. In this case the polymers are completely miscible and exist as a single, homogeneous phase. The curve of *case 3* also indicates two glass transitions but since the T_g are shifted with respect to those of the starting materials it is clear that both phases are of mixed composition. *Case 4* shows a modulus curve characteristic of a broad, ill-defined glass transition reflecting the presence of a multitude of phases each with a slightly different composition. For real polymer blends, which are attaining ever increasing commercial importance, all four cases can be observed.

In Fig. 144, as an example, the modulus/temperature curves at a constant frequency (f = 110 Hz) of SBR/BR vulcanizates of varying composition are reproduced. Only one glass transition can be identified but this is broadened and shifted towards that of BR. Thus, a spectrum of mixed phases of varying composition are present.

The modulus/temperature curve (f = 3.5 Hz) for a BR/NR vulcanizate is shown in Fig. 145. The two glass transitions suggest that these two polymers are not miscible and that two, essentially original component phases exist in the mixture.

A final example of a technically important elastomer blend is given in Fig. 146. However, in this case an unambiguous interpretation of the modulus curves is difficult since the glass transitions of the two starting elastomers NR and SBR are separated by only 10 K. Even if these two polymers were totally immiscible, separate glass transitions would probably not be visible due to the limited resolution of the method.

It is often observed that two polymers are more miscible if both are strongly polar. The modulus curves for such a system, composed of hydrogenated nitrile rubber and PVC are shown in Fig. 147. From the single glass transition exhibited by all the blends examined and the fact that the T_g reflects the composition of the blend as predicted by the Fox equation (see Eq. 130), it is clear that these two polymers are apparently miscible. Within the composition range 40/60 to 60/40 the damping maxima are, however, slightly broadened suggesting that the blend is composed of several phases of differing compositions.

Additional methods for examining the miscibility and phase behavior of polymers are: Microscopy (especially: Transmission Electron Microscopy), Light-, X-ray and Neutron Scattering, Inverse Gas-Chromatography, Calorimetry, NMR and density measurements. A detailed description of these methods can be found in [162].

Despite most polymers being immiscible with one another there are an increasing number of polymer pairs which are known to exhibit miscibility. Olabisi, Robeson and Shaw have collated over 200 such systems from the literature from which a selection is given in Table 10.

Furthermore, the miscibility of polymer mixtures can be influenced by the introduction of a third component, a compatibilizer. Due to the increasing commercial importance of polymer blends a considerable amount of experimental and theoretical research is being directed towards attaining an improved understanding of polymer miscibility and phase behavior.

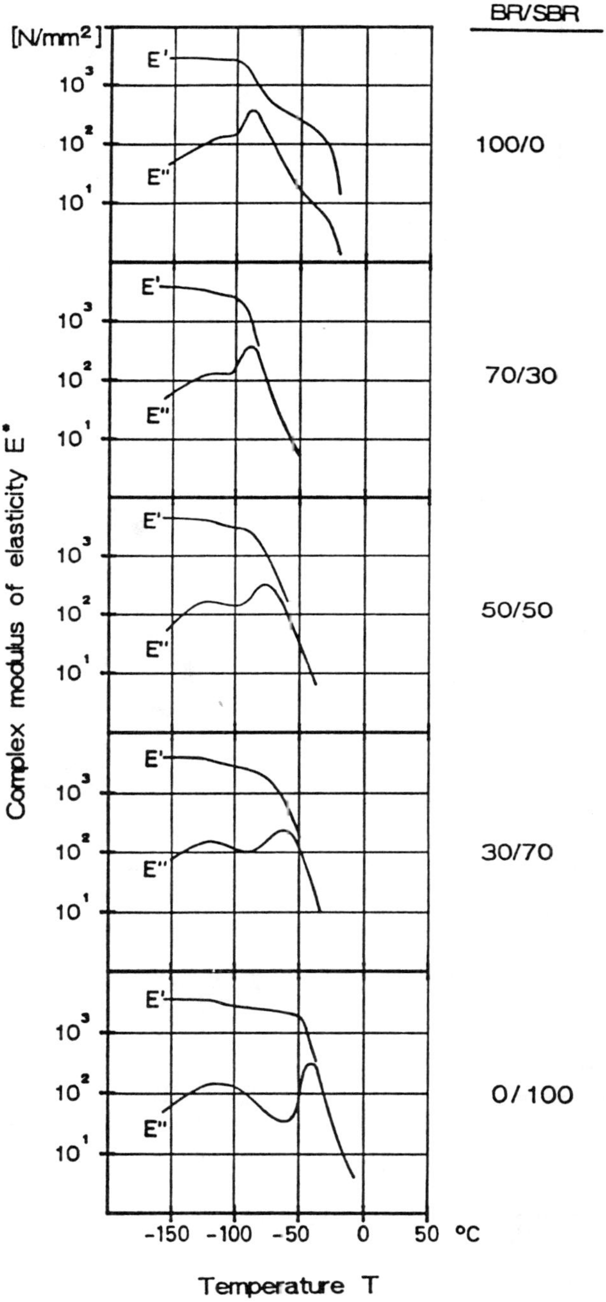

Fig. 144. Modulus/Temperature behavior of BR/SBR mixtures; measurement frequency: f = 110 Hz

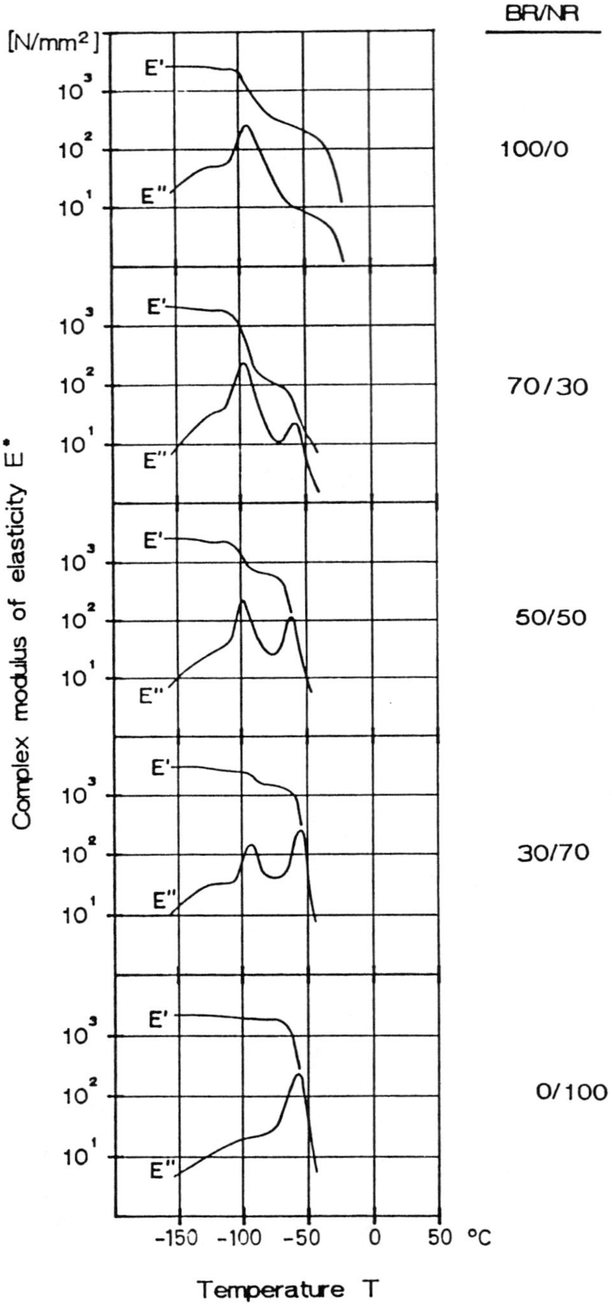

Fig. 145. Modulus/Temperature behavior of BR/NR mixtures; measurement frequency: f = 3.5 Hz

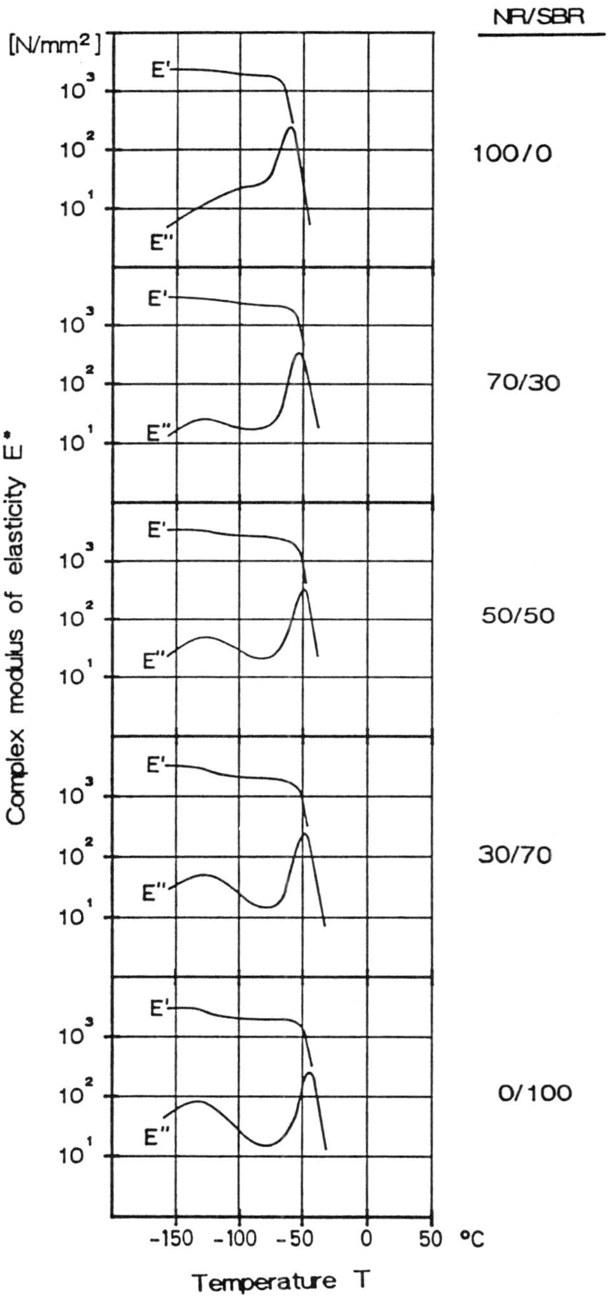

Fig. 146. Modulus/temperature behavior of NR/SBR mixtures; measurement frequency: f = 3.5 Hz

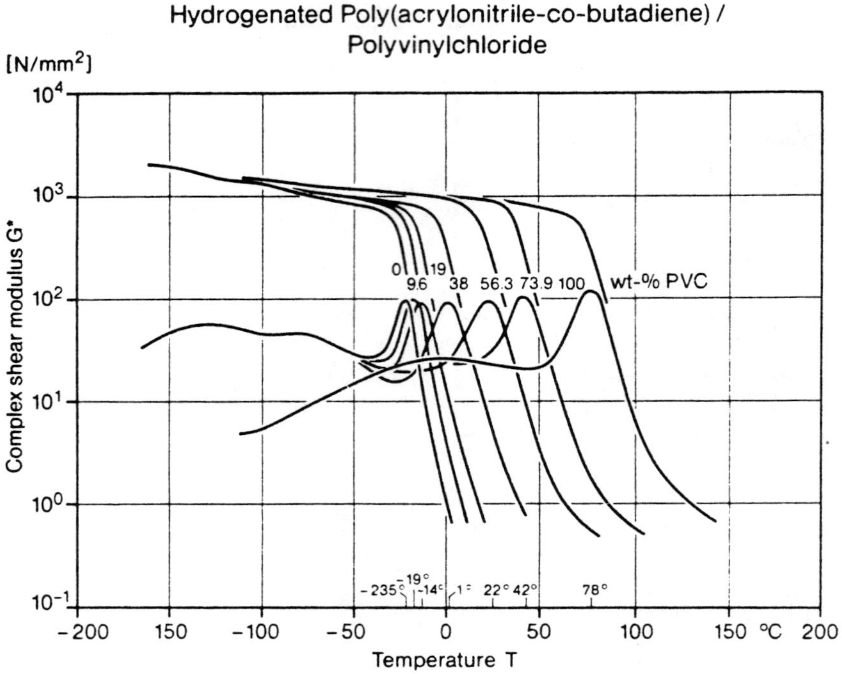

Fig. 147. Temperature dependence of the complex shear modulus for H-NBR/PVC blends

Table 10. Miscible Polymers (from [162])

Polymer components	Method	Ref.
PVC/Butadiene-co-acrylonitrile (23–45% ACN)	dynamic mechanical calorimetry microscopy	[171, 172]
PVC/Ethylene-co-vinylacetate (65–70% VA)	dynamic mechanical	[173]
PVC/Polycaprolactone	dynamic mechanical	[174]
PVC/SAN (72/28 wt. %) in PVC/ABS mixtures	dynamic mechanical	[175]
Polystyrene/PPO	dynamic mechanical calorimetry	[176, 177]
	dielectric, rheology	[178]
Polystyrene/Tetramethyl-bisphenol A polycarbonate	dynamic mechanical	[179]
Polystyrene/Polyvinyl methyl ether	calorimetry dielectric	[180]
PMMA/PVF_2	calorimetry	[181]
PMMA/Polyethylene oxide	SANS	[182]
Polycaprolactone/Polycarbonate	dynamic mechanical	[183]
Polypropylene/Polybutene-1	dynamic mechanical	[184]
PVC/Chlorinated polyethylene (42 wt. % Cl)	electron microscopy	[185]
PVC/HNBR	(see Fig. 147)	

14 Network Swelling [98, 188, 189]

The swelling of crosslinked polymers can also be theoretically described with the aid of the Flory-Huggins theory. To this end, the first step is to take account of the contribution to the free energy from the elastic deformation of the network due to the infusion of liquid molecules. The molar free energy of the swollen network is given by:

$$\Delta g_m = \Delta g_{m1} + \Delta g_{m2} \tag{355}$$

where Δg_{m1} represents the free energy of mixing for one Mole of liquid with the uncrosslinked polymer and Δg_{m2} the change in the free energy of the network due to elastic deformation during swelling. The second term can be obtained from the elastically stored energy in the network; thus, neglecting changes in internal energy:

$$W = -T\Delta S = \frac{1}{2} NkT(\lambda_1^2 + \lambda_2^2 + \lambda_3^2 - 3) \tag{356}$$

In this equation N corresponds to the number of network chains per unit volume. For an isotropic swelling ($\lambda_1 = \lambda_2 = \lambda_3 = \lambda$) in the absence of an applied stress Eq. 356 can be simplified to:

$$W = \frac{1}{2} NkT(3\lambda^2 - 3) \tag{357}$$

Putting ϕ_p equal to the volume fraction of the polymer in the swollen state one obtains for the volume of an unit cube after swelling:

$$\lambda^3 = 1/\phi_p \quad \text{or} \quad \lambda = \phi_p^{-1/3} \tag{358}$$

Substituting this into Eq. 357 gives:

$$W = \frac{3}{2} NkT(\phi_p^{-2/3} - 1) \tag{359}$$

For a swollen system containing n_s moles of swelling medium per unit volume and having a molar volume v_s, one can write λ^3 in the terms:

$$\lambda^3 = 1/\phi_p = 1 + n_s v_s \tag{360}$$

Thus, the free energy term for the deformation of the network in Eq. 355 can be written as:

$$\Delta g_{m2} = \frac{\partial W}{\partial n_s} = NkTv_s\phi_p^{1/3} \tag{361}$$

Since, $NkT = \varrho RT/M_c$ the total free energy of mixing is given by (see Eq. 343):

$$\Delta g_m = RT[\ln(1 - \phi_p) + \phi_p + \chi\phi_p^2 + (\varrho v_s/M_c)\phi_p^{1/3}] \tag{362}$$

If the swelling is continued to an equilibrium state then

$$\Delta g_m = 0 \tag{363}$$

thus, from Eq. 362:

$$\ln(1 - \phi_p) + \phi_p + \chi\phi_p^2 + (\varrho v_s/M_c)\phi_p^{1/3} = 0 \tag{364}$$

For a given χ,ϱ,v_s and M_c the equilibrium swelling is determined by ϕ_p i.e., by the volume fraction of polymer in the swollen system. For large degrees of swelling, that is, for small values of ϕ_p, Eq. 364 can be further simplified. Thus, after expanding the logarithmic term to a series and neglecting the ϕ_p terms with exponents greater than 2 one obtains:

$$\left(\chi - \frac{1}{2}\right)\phi_p^2 + (\varrho v_s/M_c)\phi_p^{1/3} \simeq 0 \tag{365}$$

or

$$(\varrho v_s/M_c) \simeq \left(\frac{1}{2} - \chi\right)\phi_p^{5/3} \tag{366}$$

This equation has been checked by Flory for a series of butyl rubber networks having variable crosslink densities using cyclohexane as swelling solvent. In addition to equilibrium swelling, an independent measurement of the stress, which is proportional to the modulus $\varrho RT/M_c$, at a given elongation was made. A double logarithmic plot of the stress versus the reciprocal of the volume fraction $(1/\phi_p)$ should, according to Eq. 366, give a straight line with a slope of 5/3. The plot for the system butyl rubber/cyclohexane (shown in Fig. 148) does indeed conform with this theoretical approach.

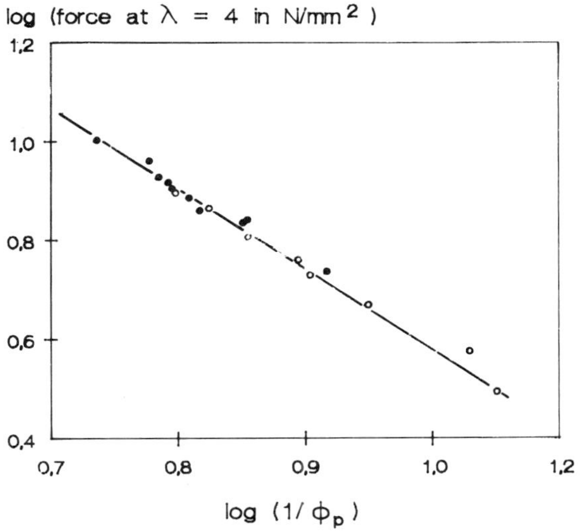

Fig. 148. Plot showing the relationship between equilibrium swelling and modulus for butyl rubber with varying crosslink densities (from [98])

The difficulties associated with a general application of Eq. 366 for the determination of crosslink densities are principally due to an insufficient knowledge of the interaction parameters. On the contrary, it is possible to determine χ from measurements of the modulus of swollen polymer networks. The modulus of a swollen network is equal to $(\varrho RT/M_c)\phi_p^{1/3}$. The stress-strain equation for a swollen network under uniaxial elongation is thus given by (see Eq. 245):

$$\sigma = (\varrho RT/M_c)\phi_p^{1/3}(\lambda - \lambda^{-2}) \tag{367}$$

This equation allows one to determine M_c at different crosslink densities with different swelling agents. Additionally, under equilibrium conditions and with

Table 11. Interaction Parameters χ for Crosslinked NR (from [98])

Swelling Agent	From Measurements of:	
	Equilibrium Swelling	Vapour Pressure
CCl_4	0.29	0.28
Chloroform	0.34	0.37
CS_2	0.425	0.49
Benzene	0.395	0.41
Toluene	0.36	0.43–0.44
Petroleum ether	0.54	0.43
n-Propyl acetate	0.62	–
Ethyl acetate	0.78	–
2-Butanone	0.94	–
Acetone	1.37	–

known values of M_c, the interaction parameter χ can be determined for each solvent from Eq. 366. A tabulation of several values of χ for NR with various solvents is given in Table 11.

From the experimental results of Gee and Mullins follows that the kinetic theory of rubber elasticity holds only for large degrees of swelling i.e., when the system is adequately close to equilibrium. This is corroborated by the fact that the Mooney-Rivlin constant C_2 (this constant is a reflection of the deviation of a material from ideal rubber elastic behavior) tends to zero with increasing degree of swelling (see Fig. 118).

15 Environmental Stress Cracking of Polymeric Materials [190]

Gases and liquids act on thermoplastics to produce an effect known as Environmental Stress Cracking (ESC). An essential characteristic of ESC is that, with an appropriate medium, damage only occurs under mechanical stress, the amount of which may be considerably less than the nominal limiting load for a given material. Even high impact thermoplastics, such as polycarbonates, behave as brittle materials under relatively low stress in the presense of some solvents (e.g. for polycarbonates: Toluene or toluene/isooctane mixtures).

L. Morbitzer et al have shown that the mechanism by which damage occurs involves a craze initiation followed by craze growth and, finally, a craze/crack transition. Craze initiation is attributed to an inhomogeneous plasticizing of the polymer by the attacking medium. This process is significantly affected by a variety of inhomogeneities. A rough indication of ESC behavior for a particular polymer in any one medium can be obtained by examining the respective solubility parameters. However, a detailed, quantitative prediction is, at the present level of understanding, not possible.

From this statement it can be assumed that in the foreseeable future there will be no alternative to the semi-empirical, predominantly qualitative approach of polymer physics, which is based on appropriate experiments. Thus, it is important that the traditional experimental methods for analyzing polymers be continually improved and adapted to keep up with technological developments. At the same time, it is equally important that the newer methods of solid state physics such as Nuclear Magnetic Resonance (NMR), Deuteron NMR, Electron Spin Resonance (ESR), Secundar Ion Mass Spectroscopy (SIMS), Laser Micro Mass Analysis (LAMMA), Neutron Scattering, Extended X-ray Absorption Fine Structure (EXAFS), X-ray Absorption Near Edge Spectroscopy (XANES), Free Volume Microprobe (FVM), etc. be increasingly applied to polymer systems.

N. W. Tschoegl, Pasadena, CA, USA

The Phenomenological Theory of Linear Viscoelastic Behavior

An Introduction

Cover illustrations: C.A. Tschoegl

1989. XXV, 769 pp. 227 figs. 25 tabs.
ISBN 3-540-19173-9

Contents:

Introductory Concepts. - Linear Viscoelastic Response. - Representation of Linear Viscoelastic Behavior by Series-Parallel Models. — Representation of Linear Viscoelastic Behavior by Spectral Response Functions. - Representation of Linear Viscoelastic Behavior by Ladder Models. - Representation of Linear Viscoelastic Behavior by Mathematical Models. - Response to Non-Standard Excitations. - Interconversion of the Linear Viscoelastic Functions. - Energy Storage and Dissipation in a Linear Viscoelastic Material. - The Modelling of Multimodal Distributions of Respondance Times. — Linear Viscoelastic Behavior in Different Modes of Deformation. - Appendix: Transformation Calculus. - Solutions to Problems. - Epilogue. - Notes on Quotation. - List of Symbols. - Author Index. - Subject Index.

Springer-Verlag
Berlin Heidelberg
New York London
Paris Tokyo
Hong Kong